Technical Bases for Yucca Mountain Standards

Committee on
Technical Bases for Yucca Mountain Standards

Board on Radioactive Waste Management

Commission on Geosciences, Environment, and Resources

National Research Council

NATIONAL ACADEMY PRESS
Washington, D.C. 1995

NOTICE: The project that is the subject of this report was approved by the Governing Board of the National Research Council, whose members are drawn from the councils of the National Academy of Sciences, the National Academy of Engineering, and the Institute of Medicine. The members of the committee responsible for the report were chosen for their special competences and with regard for appropriate balance.

This report has been reviewed by a group other than the authors according to procedures approved by a Report Review Committee consisting of members of the National Academy of Sciences, the National Academy of Engineering, and the Institute of Medicine.

Support for this study on Technical Bases for Yucca Mountain Standards was provided by the U.S. Environmental Protection Agency, under contract number 68D30014.

Library of Congress Catalog Card No. 95-69192
International Standard Book Number 0-309-05289-0

Additional copies of this report are available from:

National Academy Press
2101 Constitution Avenue, N.W.
Box 285
Washington, D.C. 20055

Call 800-624-6242 or 202-334-3313 (in the Washington Metropolitan Area).

B563

Printed in the United States of America

COMMITTEE ON TECHNICAL BASES FOR YUCCA MOUNTAIN STANDARDS

ROBERT W. FRI, *Chair,* Resources for the Future, Washington, DC,
JOHN F. AHEARNE, Sigma Xi, The Scientific Research Society, Research Triangle Park, North Carolina
JEAN M. BAHR, University of Wisconsin, Madison
R. DARRYL BANKS, World Resources Institute, Washington, DC
ROBERT J. BUDNITZ, Future Resources Associates, Berkeley, California
SOL BURSTEIN, Wisconsin Electric Power, Milwaukee (retired)
MELVIN W. CARTER, Georgia Institute of Technology, Atlanta (professor emeritus)
CHARLES FAIRHURST, University of Minnesota, Minneapolis
CHARLES McCOMBIE, National Cooperative for the Disposal of Radioactive Waste,Wettingen, Switzerland
FRED M. PHILLIPS, New Mexico Institute of Mining and Technology, Socorro
THOMAS H. PIGFORD, University of California, Berkeley, Oakland (professor emeritus)
ARTHUR C. UPTON, New Mexico School of Medicine, Santa Fe
CHRIS G. WHIPPLE, ICF Kaiser Engineers, Oakland, California
GILBERT F. WHITE, University of Colorado, Boulder
SUSAN D. WILTSHIRE, JK Research Associates, Inc., Beverly, Massachusetts

Staff

MYRON F. UMAN, Study Director
RAYMOND A. WASSEL, Senior Staff Officer
ROBERT J. CROSSGROVE, Copy Editor
LISA J. CLENDENING, Senior Project Assistant

BOARD ON RADIOACTIVE WASTE MANAGEMENT

COMMISSION ON GEOSCIENCES, ENVIRONMENT, AND RESOURCES

The National Academy of Sciences is a private, nonprofit, self-perpetuating society of distinguished scholars engaged in scientific and engineering research, dedicated to the furtherance of science and technology and to their use for the general welfare. Upon the authority of the charter granted to it by the Congress in 1863, the Academy has a mandate that requires it to advise the federal government on scientific and technical matters. Dr. Bruce Alberts is president of the National Academy of Sciences.

The National Academy of Engineering was established in 1964, under the charter of the National Academy of Sciences, as a parallel organization of outstanding engineers. It is autonomous in its administration and in the selection of its members, sharing with the National Academy of Sciences the responsibility for advising the federal government. The National Academy of Engineering also sponsors engineering programs aimed at meeting national needs, encourages education and research, and recognizes the superior achievements of engineers. Dr. Harold Liebowitz is president of the National Academy of Engineering.

The Institute of Medicine was established in 1970 by the National Academy of Sciences to secure the services of eminent members of appropriate professions in the examination of policy matters pertaining to the health of the public. The Institute acts under the responsibility given to the National Academy of Sciences by its congressional charter to be an adviser to the federal government and, upon its own initiative, to identify issues of medical care, research, and education. Dr. Kenneth Shine is president of the Institute of Medicine.

The National Research Council was organized by the National Academy of Sciences in 1916 to associate the broad community of science and technology with the Academy's purposes of furthering knowledge and of advising the federal government. Functioning in accordance with general policies determined by the Academy, the Council has become the principal operating agency of both the National Academy of Sciences and the National Academy of Engineering in providing services to the government, the public, and the scientific and engineering communities. The Council is administered jointly by both Academies and the Institute of Medicine. Dr. Bruce Alberts and Dr. Harold Liebowitz are chairman and vice chairman, respectively, of the National Research Council.

PREFACE

In Section 801 of the Energy Policy Act of 1992 (P.L. 102-486), the U.S. Congress directed the U.S. Environmental Protection Agency (EPA) to promulgate standards to ensure protection of public health from high-level radioactive wastes in a deep geologic repository that might be built under Yucca Mountain in Nevada. By this provision, EPA must set the standards to ensure protection of the health of individual members of the public. The standards will apply only to the Yucca Mountain site.

To assist EPA in this endeavor, Congress also asked the National Academy of Sciences to advise the agency on the technical bases for such standards. This report contains that advice. It was prepared by a committee organized under the auspices of the National Research Council, which is jointly managed by the National Academy of Sciences and the National Academy of Engineering for the purpose of conducting studies such as this. Biographical information on the members of our committee is presented in Appendix A.

Our charge was contained explicitly in Section 801(a)(2) of the Act and elaborated in the Conference Report accompanying the bill as well as in correspondence from the Chairman of the Senate Committee on Energy and Natural Resources, Senator J. Bennett Johnston. These documents are contained in Appendix B. The charge consisted of two parts. The first was to address three specific questions contained in Section 801(a)(2). The second was to advise EPA on the technical basis for the health-based standards that it is mandated to prepare.

To accomplish both objectives of the charge, we structured our study to focus on the state of scientific and technical understanding available for assessing the future behavior of an underground repository and for devising appropriate standards. We also took account of the eventual need for the U.S. Department of Energy (DOE) to demonstrate compliance with the standards. In the process, we conducted a series of five open technical meetings to assure that we had access to all of the analyses, including those in the international literature, that might pertain to our task. We invited to these meetings more than 50 nationally and internationally known scientists and engineers in pertinent fields to discuss the technical issues with us. At the final open meeting, we received recommendations from a number of observers about what they hoped would be in our report.

The information provided by invited experts and public participants has been most valuable to our work.

The three Federal agencies involved (the U.S. Nuclear Regulatory Commission (USNRC), in addition to DOE and EPA), state and county agencies in Nevada, and private organizations, such as the Electric Power Research Institute — all of which have sponsored research on the technical issues before us — generously shared their data and insights with us. We also retained two consultants, Paul Dejonghe (Nuclear Energy Research Center, Mol, Belgium (retired)) summarized for us the experience of other countries in setting standards for high-level radioactive waste repositories, and Detlof von Winterfeldt (University of Southern California, Los Angeles) reviewed the literature on human intrusion into a repository at some time in the future and institutional controls to mitigate such intrusions. All of the information we received is available to the public.

Although we believe that the full range of scientific information related to the standards was available to us, it became clear in the course of our work that designing the standards requires making decisions based as much or more on policy considerations than on science. It is equally clear that there is no sharp dividing line between science and policy. In developing this report, we have recognized that the committee members can speak as experts only on matters of science, but we have not construed our assignment so narrowly as to limit the usefulness of our recommendations for standard-setting in the real world. In short, we have commented on policy issues where we thought it necessary.

Science alone cannot answer policy questions, however, and so we do not make policy recommendations in this report on the grounds that there is by definition a limited scientific basis for selecting one policy alternative over another. We have instead tried to use available technical information and judgment to suggest a starting point for the rulemaking process that will lead to a policy decision.

By applying this approach consistently throughout the report, we have achieved consensus on a number of complex and controversial issues. In particular, we agree on the answers to the questions posed by Congress, the technical bases for health-based standards, and most of the elements of a procedure for assessing compliance with the standards. On one aspect of the compliance assessment procedure, however, one member of the committee has prepared a personal statement presented in Appendix E, outlining where his views diverge from the view of the rest of the committee. As chair of the committee, I have provided my perspective on that statement in Appendix F.

The issue in question concerns what assumptions to make about the future distribution of individuals that would be exposed to any possible releases from the repository. Assessing compliance, which would be required in any licensing procedure, requires that the populations or individuals at risk be specified so their exposures or risks can be estimated and compared with the standards. Because the population at greatest risk from repository releases will exist far in the future, that population and its distribution are not amenable to scientific prediction. Developing a complete exposure scenario for compliance analysis therefore requires making assumptions about who will be at risk. While scientific information can be valuable in developing these assumptions, the choice of assumptions is ultimately a matter for policy judgment.

In the committee's view, there are two avenues for approaching the construction of exposure scenarios. The two are summarized and compared in Chapter 3 and described in detail in Appendixes C and D, respectively. The committee offers these approaches as options for regulators to consider in specifying an exposure scenario. The personal statement by one committee member in Appendix E argues that only one of these approaches is appropriate. Although others may have an equally strong preference for the other approach, the remainder of the committee has preferred to follow its consistent practice of not taking a position on policy questions.

On behalf of my colleagues on the committee, I wish to express our appreciation to all those who provided us with valuable input for our task. In particular, at our request, several agencies formally designated liaisons to the committee whose assignments were to assure that we were fully informed of the data and analyses that were available to their respective agencies. In turn, through the liaisons, we shared all of the information we obtained during the course of our work with their agencies. The liaisons were J. William Gunter, EPA; Margaret Federline, USNRC; Stephen J. Brocoum, DOE; Les W. Bradshaw, Nye County (Nevada) Nuclear Waste Repository Program; and Engelbrecht von Tiesenhausen, Clark County, Nevada. Robert Loux performed a similar function, although informally, for the State of Nevada Nuclear Waste Projects Office. Rosa Yang made sure that we were fully informed of the results of research performed under the auspices of the Electric Power Research Institute. We are indebted to these individuals for their dedication to providing us with information.

Finally, our work, while difficult enough, would have been even more so without the dedicated support of the National Research Council staff: Ray Wassel, Myron Uman, Lisa Clendening, and others who also assisted the committee.

Robert W. Fri
Chair

CONTENTS

List of Figures

List of Tables

EXECUTIVE SUMMARY

Proper management of high-level radioactive wastes, including those resulting from the production of nuclear weapons and the operation of nuclear electric power plants, is vital for the protection of the public health and safety. It has been longstanding federal policy to dispose of these wastes underground in a mined geologic repository. The U.S. Department of Energy (DOE) is charged with the development and eventual operation of a repository. The U.S. Environmental Protection Agency (EPA) and the U.S. Nuclear Regulatory Commission (USNRC) share the responsibility for regulating the disposal program to ensure adequate protection of the health and safety of the public.

EPA promulgated its first standard for deep geologic disposal of high-level radioactive waste in 1985; this standard was challenged, litigated, and ultimately reissued in 40 CFR 191 in December 1993. Before EPA promulgated the new standard, however, Congress enacted the Energy Policy Act of 1992, which mandated a separate process for setting a standard specifically for the proposed repository at Yucca Mountain, Nevada. In Section 801 of the Act, Congress required EPA to arrange for an analysis by the National Academy of Sciences of the scientific basis for a standard to be applied at the Yucca Mountain site and directed EPA," based upon and consistent with the finding and recommendations of the National Academy of Sciences, [to] promulgate, by rule, public health and safety standards for protection of the public from releases from radioactive materials stored in or disposed of in the repository at the Yucca Mountain site." This report responds to the charge of Section 801.

Implicit in setting a Yucca Mountain standard, is the assumption that EPA, USNRC, and DOE can, with some degree of confidence, assess the future performance of a repository system for time scales that are so long that experimental methods cannot be used to confirm directly predictions of the behavior of the system or even of its components. This premise raises the basic issue of whether scientifically justifiable analyses of repository behavior over many thousands of years in the future can be made. We conclude that such analyses are possible, within restrictions noted in this report. Nevertheless, these assessments of repository performance must contend with substantial uncertainties, and some areas

— projecting the behavior of human society over very long periods, for example — are beyond the limits of scientific analysis. We have made explicit those instances, and have also pointed out where we believe it is appropriate to rely on informed judgments and reasonable assumptions to supplement scientific analysis.

In attempting to make the best use of the scientific understanding that is available, we have arrived at recommendations that differ in important ways from the approach followed by EPA in 40 CFR 191. In particular, we recommend:

- The use of a standard that sets a limit on the risk to individuals of adverse health effects from releases from the repository. 40 CFR 191 contains an individual-dose standard, and it continues to rely on a containment requirement that limits the releases of radionuclides to the accessible environment. The stated goal of the containment requirement was to limit the number of health effects to the global population to 1,000 incremental fatalities over 10,000 years. We do not recommend that a release limit be adopted.
- That compliance with the standard be measured at the time of peak risk, whenever it occurs.[1] The standard in 40 CFR 191 applies for a period of 10,000 years. Based on performance assessment calculations provided to us, it appears that peak risks might occur tens to hundreds of thousands of years or even farther into the future.
- Against a risk-based calculation of the adverse effect of human intrusion into the repository. Under 40 CFR 191, an assessment must be made of the frequency and consequences of human intrusion for purposes of demonstrating compliance with containment requirements. In contrast, we conclude that it is not possible to assess the frequency of intrusion far into the future. We do recommend that the consequences of an intrusion be calculated to assess the resilience of the repository to intrusion.

---

[1] Within the limits imposed by the long-term stability of the geologic environment, which is on the order of one million years.

Finally, we have identified several instances where science cannot provide all of the guidance necessary to resolve an issue. This is particularly true in developing procedures for compliance assessment. Setting the standard, therefore, requires addressing policy questions as well as scientific ones. We recommend that resolution of policy issues be done through a rulemaking process that allows opportunity for wide-ranging input from all interested parties. In these cases, we have tried to suggest positions that could be used by the responsible agency in formulating a proposed rule. Other starting positions are possible, and of course the final rule could differ markedly from any of them.

Although we have taken a broad view of the scientific basis for the standard, we have not addressed the social, political, and economic issues that might have more effect on the repository program than the health standard. In particular, we have not recommended what levels of risk are acceptable; we have not considered whether the development of a permanent repository should proceed at this time; nor have we made a judgment about the potential for the Yucca Mountain site to comply with the standard eventually adopted.

PROTECTING HUMAN HEALTH

In Section 801, Congress directs that EPA set a standard for Yucca Mountain by specifying the maximum annual effective dose equivalent to individual members of the public. The first question posed in Section 801 is whether such a standard will provide a reasonable basis for protecting the health and safety of the general public. We recommend the use of a standard designed to limit individual risk, and describe how a standard might be structured on this basis. We then address the specific question of protection of public health in the context of the individual-risk standard and compare this standard to the one currently used by EPA. Based on this analysis, we conclude not only that the individual risk standard would protect the health of the general public, but also that it is a particularly appropriate standard for the Yucca Mountain site in light of the characteristics of this site.

The risks to humans from exposures to low levels of radiation have been assessed in detail by national and international organizations. These assessments are fraught with uncertainty, but it has been possible to reach a reasonable consensus within the scientific community on the relationship of dose and health effects, which is generally considered to provide an acceptable basis for evaluating the risks attributable to a given dose or the degree of protection afforded by a given limitation of exposure. Additionally, a general consensus exists among national and international bodies on a framework for protecting the public health that provides a limit of 1 milliSievert (mSv) (100 millirem (mrem)) per year effective dose for continuous or frequent exposures from all anthropogenic sources of ionizing radiation other than medical exposures. A general consensus also appears to exist among national authorities in various countries to accept and use the principle of apportioning this total radiation dose limit among the respective anthropogenic sources of exposure, typically allocating to high-level waste disposal a range of 0.1 to 0.3 mSv (10 to 30 mrem) per year.

Elements of the Standard

A standard is a societally acceptable limit on some aspect of repository performance that should not be exceeded if the repository is to be judged safe. We recommend the use of a standard that sets a limit on the risk to individuals of adverse health effects from releases from the repository. A risk-based standard would not have to be revised in subsequent rulemaking if advances in scientific knowledge reveal that the dose-response relationship is different from that envisaged today. Such changes have occurred frequently in the past, and can be expected to occur in the future. For example, ongoing revisions in estimates of the radiation doses received by atomic bomb survivors of Hiroshima and Nagasaki might significantly modify the apparent dose-response relationships for carcinogenic effects in this population, as have previous revisions in dosimetry (see Straume et al., 1992). Moreover, risks to human health from different sources, such as nuclear power plants and toxic chemicals can be compared in reasonably understandable terms.

It is essential to define specifically how to calculate risk, however, for otherwise it will not be clear what number to use to compare to the risk limit established in the standard. We define risk as the expected value of a probabilistic distribution of health effects. The first step in calculating risk is to develop a distribution of doses received by individuals. A probabilistic distribution of health effects can be developed as the product of each value of dose received and the health effect per unit dose.

Structuring of the individual-risk standard requires specifying what level of protection is to be afforded, who is to be protected, and for how long. We acknowledge that determining what risk level is acceptable is not ultimately a question of science but of public policy. We note, however, that EPA has already used a dose limit equivalent to a risk level of 5×10^{-4} health effects in an average lifetime, or a little less than 10^{-5} effects per year assuming an average lifetime of 70 years, as an acceptable risk limit in its recently published 40 CFR 191. This limit is consistent with limits established by other federal nuclear regulations. In addition, the risk equivalent of the dose limits set by authorities outside the United States is also in the range of 10^{-5} to 10^{-6} per year (except for exposure to radon indoors or releases from mill tailings). This range is a reasonable starting point for EPA's rulemaking.

To determine whether a repository complies with the standard, it is necessary to calculate the risk to some individual or representative group of individuals and then to compare the result to the risk limit established in the standard. Therefore, the standard must specify the individual or individuals for whom the risk calculation is to be made. Although not strictly a scientific issue, we believe that the appropriate objective is to protect the vast majority of members of the public while also ensuring that the decision on the acceptability of a repository is not unduly influenced by the risks imposed on a very small number of individuals with unusual habits or sensitivities. The situation to be avoided, therefore, is an extreme case defined by unreasonable assumptions regarding the factors affecting dose and risk, while meeting the objectives of protecting the vast majority of the public. An approach that is consistent with this objective, and is used extensively elsewhere in the world, is the critical-group approach. <u>We recommend that the critical-group approach be used in the Yucca Mountain standards</u>.

The critical group has been defined by the International Commission on Radiological Protection (ICRP) as a relatively homogeneous group of people whose location and habits are such that they are representative of those individuals expected to receive the highest doses[2] as a result of the discharges of radionuclides. Therefore, as the ICRP notes, "because the actual doses in the entire population will constitute a distribution for which the critical group represents the extreme, this procedure is intended to ensure that no individual doses are unacceptably high." (ICRP, 1985a, at paragraph 46). In the context of an individual-risk standard, and using cautious, but reasonable, assumptions, the group would include the persons expected to be at highest risk, would be homogeneous in risk[3], and would be small in number. The critical-group risk calculated for purposes of comparison with the risk limit established in the standard would be the mean of the risks to the members of the group.

This definition requires specifying the persons who are likely to be at highest risk. In the present and near future, these persons are real; that is, they are the persons now living in the near vicinity of the repository and in the direction of the postulated flow of the plume of radionuclides. For the far future, however, it will be necessary to define hypothetical persons by making assumptions about lifestyle, location, eating habits, and other factors. The ICRP recommends use of present knowledge and cautious, but reasonable, assumptions.

The current EPA standard contains a time limit of 10,000 years for the purpose of assessing compliance. We find that there is no scientific basis for limiting the time period of an individual-risk standard in this way. We believe that compliance assessment is feasible for most physical and geologic aspects of repository performance on the time scale of the long-term stability of the fundamental geologic regime — a time scale that is on the order of 10^6 years at Yucca Mountain — and that at least some potentially important exposures might not occur until after several hundred thousand years. For these reasons, we recommend that compliance

---

[2] The ICRP defines critical group in dose terms. We use the ICRP terminology here to describe the concept as developed by the ICRP, and later adapt the concept to the risk framework.

[3] That is, the difference between the highest and lowest risk faced by individuals in the group should be relatively small. Should a radiation dose occur, however, it may affect only a few members of the group. This is the difference between risk (the chance of an adverse health effect) and outcome (a cancer that actually develops). Risk can be homogeneous, even when outcomes are quite diverse.

assessment be conducted for the time when the greatest risk occurs, within the limits imposed by long-term stability of the geologic environment.

Another time-related regulatory concern, based on ethical principles, is that of intergenerational equity. A health-based risk standard could be specified to apply uniformly over time and generations. Such an approach would be consistent with the principle of intergenerational equity that requires that the risks to future generations be no greater than the risks that would be accepted today. Whether to adopt this or some other expression of the principle of intergenerational equity is a matter for social judgment.

Protection of the General Public

Congress has asked whether a standard intended to protect individuals would also protect the general public in the case of Yucca Mountain. We conclude that an individual-risk standard would protect public health, given the particular characteristics of the site, provided that policy makers and the public are prepared to accept that very low radiation doses pose a negligibly small risk.

The individual risk-standard that we recommend is intended to protect a critical group. In this context, the general public includes both global populations as well as local populations that lie outside the critical group. Global populations might be affected because radionuclide releases from a repository can in theory be diffused throughout a very large and dispersed population. In the case of Yucca Mountain, the likely pathway leading to widely dispersed radionuclides is via the atmosphere beginning with release of carbon dioxide gas containing the carbon-14 (^{14}C) radioactive isotope which might escape from the waste canisters.

The risks of radiation produced by such wide, dispersion are likely to be several orders of magnitude below those of a local critical group. Great uncertainty exists about the number of health effects that would be imposed on the global population because of the difficulties in interpreting the risks associated with very small incremental doses of radiation. As noted in the BEIR V report (NRC, 1990a), the lower limit of the range of uncertainty in such risk estimates extends to zero (no effects). To address scenarios of widespread but extremely low-level doses, the radiation protection community has introduced the concept of negligible incremental dose (above background levels). For example, the National Council on Radiation Protection and Measurements (NCRP) has recommended a value of 0.01 mSv/yr (1mrem/yr) per radiation source or practice (NCRP 1993), which currently would correspond to a projected risk of about 5×10^{-7}/yr for fatal cancers, assuming the linear hypothesis. We believe that this concept can be extended to risk and can be applied to the establishment of a radiation standard at Yucca Mountain. Defining the level of incremental risk that is negligible is a policy judgment. We suggest the risk equivalent of the negligible individual incremental dose recommended by the NCRP as a reasonable starting point for developing consensus.

Persons in some population outside the critical group may, however, still be exposed to risks in excess of the level of the negligible incremental risk but below the level of the critical group risk. The risks to these persons as individuals are, by definition, acceptable, but whether the effects on this population as a whole are acceptable remains a matter of judgment. Based on our review, we conclude that there is no technical basis for a population risk standard by which to make such a judgment.

ASSESSING COMPLIANCE

Any standard to protect individuals and the public after the proposed repository is closed will require assessments of performance at times so far in the future that a direct demonstration of compliance is out of the question. The only way to evaluate the risks of adverse health effects and to compare them with the standard is to assess the estimated potential future behavior of the entire repository system and its potential effects on humans. This procedure, involving modeling of processes and events that might lead to releases and exposures, is called performance assessment.

The technical feasibility of developing performance assessment calculations to evaluate compliance with a risk standard at Yucca Mountain depends on the feasibility of modeling the relevant events and processes (including their probabilities) specific to that site. By soliciting technical appraisals at our open meetings, reviewing solicited and unsolicited written contributions, and drawing on the available literature and our own experience and expertise, we have assessed the types, magnitudes, and time-dependencies of the uncertainties associated with potential radionuclide transport from a Yucca Mountain repository, the effects of potential natural and human modifiers of repository performance, and the pathways through the biosphere.

Physical and Geologic Processes

The properties and processes leading to transport of radionuclides away from the repository include release from the waste form, transport to the near-field zone, gas phase transport to the atmosphere above Yucca Mountain and its dispersal in the world atmosphere, and transport from the unsaturated zone to the water table and from the aquifer beneath the repository to other locations from which water might be extracted by humans. <u>We conclude that these physical and geologic processes are sufficiently quantifiable and the related uncertainties sufficiently boundable that the performance can be assessed over time frames during which the geologic system is relatively stable or varies in a boundable manner</u>. The geologic record suggests that this time frame is on the order of 10^6 years. We further conclude that the probabilities and consequences of modifications by climate change, seismic activity, and volcanic eruptions at Yucca Mountain are sufficiently boundable that these factors can be included in performance assessments that extend over this time frame.

Exposure Scenarios

Performance assessment of physical and geologic processes will produce estimates of potential concentrations of radionuclides in ground water or air at different locations and times in the future. To proceed from these concentrations to calculations of risks to a critical group requires the development of an exposure scenario that specifies the pathways by which persons would be exposed to radionuclides released from the repository. Once an exposure scenario has been adopted, performance assessment calculations can be carried out with a degree of uncertainty comparable to the uncertainty associated with geologic processes and engineered systems.

<u>Based upon our review of the literature, we conclude, however, that it is not possible to predict on the basis of scientific analyses the</u>

societal factors required for an exposure scenario. Specifying exposure scenarios therefore requires a policy decision that is appropriately made in a rulemaking process conducted by EPA. We recommend against placing the burden of postulating and defending an exposure scenario on the applicant for the license.

As with other aspects of defining standards and demonstrating compliance that involve scientific knowledge but must ultimately rest on policy judgments, we considered what to suggest to EPA as a useful starting point for rulemaking on exposure scenarios. Reflecting the disagreement inherent in the literature, we have not reached complete consensus on this question. It is essential that the scenario that is ultimately selected be consistent with the critical-group concept that we have advanced. Additionally, EPA should rely on the guidance of ICRP that the critical group be defined using present-day knowledge with cautious, but reasonable, assumptions.

We considered two illustrative approaches to the design of an exposure scenario that EPA might propose to initiate the rulemaking process. The approaches have many elements in common but differ in their treatment of assumptions about the location and lifestyle of persons who might be exposed to releases from the repository, and in the method of calculating the average risk of the members of the critical group. A substantial majority of the committee members, but not all, considers one of the approaches to be more consistent with the foregoing criteria. This particular approach explicitly accounts for how the physical characteristics of the site might influence population distribution and identifies the makeup of the critical group probabilistically.

HUMAN INTRUSION

Human activity that penetrates the repository (by drilling directly into it from the surface, for example) can cause or accelerate the release of radionuclides. Waste material could be brought to the surface and expose the intruder to high radiation doses, or the material could disperse into the biosphere. The second and third questions asked in Section 801 of the Energy Policy Act of 1992 concern the potential that at some time people might intrude into the repository.

With respect to the second question of Section 801, we conclude that it is not reasonable to assume that a system for post-closure oversight of the repository can be developed, based on active institutional controls, that will prevent an unreasonable risk of breaching the repository's engineered barriers or increasing the exposure of individual members of the public to radiation beyond allowable limits. This conclusion is founded on the absence of any scientific basis for making projections over the long term of the social, institutional, or technological status of future societies. Additionally, there is no technical basis for making forecasts about the long-term reliability of passive institutional controls, such as markers, monuments, and records.

With respect to the third question in Section 801, we conclude that it is not possible to make scientifically supportable predictions of the probability that a repository's engineered or geologic barriers will be breached as a result of human intrusion over a period of 10,000 years. We reach this conclusion because we cannot predict the probability that a future intrusion would occur in a given future time period or the probability that a future intrusion would be detected and remediated, either when it occurs or later. In addition, we cannot predict which resources will be discovered or will become valuable enough to be the objective of an intruder's activity. We cannot predict the characteristics of future technologies for resource exploration and extraction, although continued developments in current noninvasive geophysical techniques could substantially reduce the frequency of exploratory boreholes.

Although there is no scientific basis for judging whether active institutional controls can prevent an unreasonable risk of human intrusion, we think that, if the repository is built, such controls and other activities might be helpful in reducing the risk of intrusion, at least for some initial period of time after a repository is closed. Therefore, we believe that a collection of prescriptive requirements, including active institutional controls, record-keeping, and passive barriers and markers would help to reduce the risk of human intrusion, at least in the near term.

Moreover, because it is not technically feasible to assess the probability of human intrusion into a repository over the long term, we do not believe that it is scientifically justified to incorporate alternative scenarios of human intrusion into a fully risk-based compliance assessment. We do, however, conclude that it is possible to carry out calculations of the consequences for particular types of intrusion events.

The key performance issue is whether repository performance would be substantially degraded as a consequence of an intrusion of the type postulated. For this purpose, we have focused on the particular class of cases in which the intrusion is inadvertent and the intruder does not recognize that a hazardous situation has been created.

To provide for the broadest consideration of what human intrusion scenario or scenarios might be most appropriate, we recommend that EPA make this determination in its rulemaking to adopt a standard. For simplicity, we considered a stylized intrusion scenario consisting of one borehole of a specified diameter drilled from the surface through a canister of waste to the underlying aquifer. In our view, the performance of the repository, having been intruded upon, should be assessed using the same analytical methods and assumptions, including those about the biosphere and critical groups, used in the assessment of performance for the undisturbed case. We recommend that EPA require that the estimated risk calculated from the assumed intrusion scenario be no greater than the risk limit adopted for the undisturbed-repository case because a repository that is suitable for safe long-term disposal should be able to continue to provide acceptable waste isolation after some type of intrusion. As with other policy-related aspects of our recommendations, we note that EPA might decide that some other risk level is appropriate.

IMPLICATIONS OF OUR CONCLUSIONS

Limits of the Scientific Basis

It might be possible that some of the current gaps in scientific knowledge and uncertainties that we have identified might be reduced by future research. It seems reasonable, therefore, to ask what gaps could be closed by taking time to obtain more scientific and technical knowledge on such matters as the nature of the waste, its potential use, the health effects of radionuclides, the value of waste products for later generations, and the security of retrievable storage containers. New information in these and other areas could improve the basis for setting the standards.

Whether the benefit of new information would be worth the additional time and resources required to obtain it is a matter of judgment. This judgment would be strengthened by a careful appraisal of the probable costs and risks of continuing the present temporary waste disposal practices and storage facilities as compared to those attaching to the proposed repository. No such comprehensive appraisal is now available. Conducting such an appraisal, however, should not be seen as a reason to slow down ongoing research and development programs, including geologic site characterization, or the process of establishing a standard to protect public health.

Technology-Based Standards

Technology-based standards play an important role in regulations designed to protect the public health from the risks associated with nuclear facilities. We have examined three technological approaches in our study.

The "as low as reasonably achievable" (ALARA) principle is intended to be applied after threshold regulatory requirements have been met, and calls for additional measures to be taken to achieve further reduction in the calculated health effects. While ALARA continues to be widely recommended as a philosophically desirable goal, its applicability to geologic disposal of high-level waste is limited at best because the technological alternatives available for designing a geologic repository are quite limited. Further, the difficulties of demonstrating technical or legal compliance with any such requirement for the post-closure phase could well prove insuperable even if it were restricted to engineering and design issues. We conclude that there is no scientific basis for incorporating the ALARA principle into the EPA standard or USNRC regulations for the repository.

If EPA issues standards based on individual risk, the USNRC would be required to revise its current regulations embodied in 10 CFR 60 to be consistent with such standards. One purpose of 10 CFR 60, which contains technology specifications, is to help ensure multiple barriers within the repository system. We conclude that because it is the performance of the total system in light of the risk-based standard that is crucial, imposing subsystem performance requirements might result in suboptimal repository design.

Finally, several persons suggested to our committee the use of a technology-based standard that would specify a strict release limit from an engineered barrier system during the early life of the repository. We find that such a limitation on early releases would have no effect on the results of compliance analysis over the long-term. Nonetheless some members of the committee believe that such a limitation might provide added assurance of safety in the near-term, and EPA might wish to consider this as a policy matter.

Administrative Consequences

Our recommendations, if adopted, imply the development of regulatory and analytical approaches for Yucca Mountain that are different from those employed in the past and from some approaches currently used elsewhere by EPA. The change in approach and the time required to develop a thorough and consistent regulatory proposal and to provide for full public participation in the rulemaking process will require considerable effort by EPA. This process probably will take more than the year, currently provided in statute, for EPA to complete development of a Yucca Mountain standard in a technically competent way. This does not mean that DOE's Yucca Mountain Site Characterization Project cannot proceed usefully in the interim.

1

INTRODUCTION

Proper management of high-level radioactive wastes, including those resulting from the production of nuclear weapons and the operation of nuclear electric power plants, is vital for the protection of public health and safety. In the United States, defense wastes from the nuclear weapons program have been accumulating for about 50 years and spent nuclear fuel from commercial power plants has been accumulating for almost 40 years.

Together defense nuclear wastes and spent nuclear fuel have been generated at almost 100 sites located throughout the country. At present, high-level defense wastes are in various physical and chemical forms and are stored—much of it in underground steel tanks—in several types of facilities, primarily at three U.S. Department of Energy (DOE) weapons-complex locations: Hanford site, WA; Savannah River site, SC; and the Idaho National Engineering Laboratory, ID (DOE, 1993a). The commercial spent nuclear fuel is stored in water pools and in above-ground dry-storage casks at more than 70 sites throughout the U.S.

There is therefore a need for a long-term strategy for disposal of these wastes that limits to an acceptable level the risks that they pose to public health and safety. By law, providing for "permanent disposal" of high-level radioactive waste is the responsibility of the federal government. It has been longstanding federal policy (see the Nuclear Waste Policy Act of 1982 (P.L. 97-425)) to dispose of these wastes in an underground mined geologic repository; the geologic disposal option has been examined and generally endorsed by the scientific community (National Research Council (NRC),1957, 1983, 1990b).

The responsibility for high-level radioactive waste disposal is divided among three federal agencies. DOE is charged with the development and eventual operation of a geologic repository. It must locate an appropriate site; demonstrate the site's ability to meet regulatory requirements; obtain a license from the U.S. Nuclear Regulatory Commission (USNRC); and construct, operate, and maintain surveillance of the repository itself. The U.S. Environmental Protection Agency (EPA) and the USNRC share the responsibility for regulating the disposal program to ensure adequate protection of the health and safety of the public. Operating under the authority of the Atomic Energy Act of 1954 (42 USC 2201(b)), EPA must establish generally applicable standards for protection of the environment from offsite releases from radioactive material in repositories (see 42 USC 1014(a), and the Nuclear Waste Policy Act of 1982 (P.L. 97-425)). The USNRC promulgates technical regulations that are consistent with the standards and considers license applications from DOE for any proposed repository, determining with reasonable assurance whether the EPA standard can be met. USNRC will have continued regulatory responsibilities to oversee the repository operation.

The process of selecting a deep geologic repository for high-level radioactive waste in the United States has been going on since at least 1975, although DOE has yet to apply for a license to build such a repository. In 1987, Congress directed DOE's Office of Civilian Radioactive Waste Management to concentrate only on the Yucca Mountain Site (Nuclear Waste Policy Act Amendments of 1987). DOE is currently studying the Yucca Mountain site by a process called "site characterization" to accumulate the information necessary to judge whether it will meet the standard to be set by EPA. If the site is deemed appropriate to be considered in the licensing process and a license application to USNRC is approved, DOE estimates that the earliest date for possible emplacement of high-level radioactive waste at Yucca Mountain would be the year 2010 (C. Gertz, DOE, personal communication, May 28, 1993). If the site is not deemed appropriate, Congress requires, in Section 113 of the Nuclear Waste Policy Act, recommendations from the Secretary of DOE to assure the safe, permanent disposal of spent nuclear fuel and high-level radioactive waste, including the need for new legislative authority.

This report deals with only one aspect of this long and complicated process — the standard that must be set to protect public health. The standard-setting process itself has extended over a period of nearly twenty years. EPA promulgated its first standard for deep geologic disposal of high-level radioactive waste (40 CFR 191) in 1985, after about a decade of study. Consistent with the directive of its authorizing statute, EPA intended this standard to be generally applicable to any deep geologic disposal site. At the time, several repository sites were being considered for spent nuclear fuel and defense high-level waste, and the Waste Isolation

Pilot Plant (WIPP) near Carlsbad, New Mexico, was being designed to accept transuranic waste from the defense nuclear program.[1]

Challenged by intervenors and state agencies, the standard was judicially reviewed, and in 1987 the U.S. Court of Appeals for the First Circuit remanded the standard to EPA for reconsideration of several of its provisions. Before EPA promulgated a new standard, however, Congress enacted the Energy Policy Act of 1992 (P.L. 102-486), which mandated a separate process for setting a standard specifically for the proposed repository at Yucca Mountain, Nevada. Through Section 801 of the Act, Congress severed the Yucca Mountain standard from coverage under the generally applicable standard in 40 CFR 191 and the Atomic Energy Act of 1954. In December 1993, EPA issued a final regulation (as 40 CFR 191) responding to the issues raised in the 1987 court remand, but this revised regulation does not apply to the proposed repository at Yucca Mountain.

In Section 801, Congress mandated that EPA arrange for an analysis by the National Academy of Sciences of the scientific basis for standards to be applied at the Yucca Mountain site and directed the agency," based upon and consistent with the findings and recommendations of the National Academy of Sciences, [to] promulgate, by rule, public health and safety standards for protection of the public from releases from radioactive materials stored in or disposed of in the repository at the Yucca Mountain site." The first paragraph of Section 801(a) provides that the standard prescribe the maximum annual effective dose equivalent to individual members of the public from releases to the accessible environment. These standards will be the only ones for high-level radioactive waste disposal applicable to the Yucca Mountain site, and are to be promulgated within one year after the Academy submits its study. USNRC then has one year to issue its specific regulations, requirements, and criteria to be consistent with the EPA Yucca Mountain standard.

This report responds to the charge made explicit in Section 801(a)(2), and in particular to the three questions that it posed:

---

[1] According to the definition provided in 40 CFR 191, "transuranic waste" is waste that is contaminated with alpha-emitting radionuclides with atomic numbers greater than that of uranium (92), half-lives greater than 20 years, and concentrations greater than 1 ten-millionth of a curie per gram of waste.

1. Whether a health-based standard based upon doses to individual members of the public from releases to the accessible environment . . . will provide a reasonable standard for the protection of the health and safety of the general public.
2. Whether it is reasonable to assume that a system for post-closure oversight of the repository can be developed, based upon active institutional controls, that will prevent an unreasonable risk of breaching the repository's engineered barriers or increasing the exposure of individual members of the public to radiation beyond allowable limits.
3. Whether it is possible to make scientifically supportable predictions of the probability that a repository's engineered or geologic barriers will be breached as a result of human intrusion over a period of 10,000 years.

The conference report accompanying Section 801 makes clear that Congress does not intend for our report to "establish specific standards for protection of the public but rather to provide expert scientific guidance on the issues involved in establishing those standards." (See Congressional Record, Oct. 8, 1992, pp. S17555 and H11399.) Furthermore, the conference report and subsequent correspondence, dated May 20, 1993, from the Chairman of the Senate Energy and Natural Resources Committee point out that our study is not precluded from addressing additional issues. (See Appendix B for the language of P.L. 102-486, the accompanying conference report, and the correspondence.) Accordingly, the scope of this report embraces a range of scientific questions about the Yucca Mountain standards and the process of demonstrating compliance with the standard.

SCOPE OF THE STUDY

The disposal of high-level radioactive waste in a geologic repository initially requires placing radionuclides in the repository at concentrations far in excess of natural levels. Some radionuclides decay quickly: for example cesium-137 has a half-life of 30 years and strontium-90 has a half-life of about 29 years. But some of the radionuclides have long half-lives: for example, the half-life of carbon-14 is 5,730 years and the half-life of iodine-129 is 17 million years. Others produce decay products that in turn persist for very long periods. The half-lives of plutonium-239 and neptunium-237 are 24,360 years and 2.2 million years, respectively.

The purpose of deep geologic disposal is to provide long-term barriers to the escape of these radionuclides into the biosphere[2]. Most of the original radioactive material placed in a repository is expected to have decayed to natural background levels while these barriers are effective. However, some of the longer-lived radionuclides involved will ultimately enter the biosphere, although it might take tens to hundreds of thousands of years or longer to do so. These releases will be "acceptable" in a regulatory sense if the adverse consequences for public health are sufficiently low. The health standard to be set by EPA and compliance with the standard will, in principle, determine whether the residual risks are acceptable.

Implicit in setting such a standard, and in demonstrating compliance with it, is the assumption that EPA, USNRC, and DOE can, with some degree of confidence, assess the future performance of a repository system for time scales that are so long that experimental methods cannot be used to confirm directly predictions of the behavior of the system or even of its components. This premise raises the basic issue of whether scientifically justifiable analyses of repository behavior over many thousands of years in the future can be made. Based on our evaluation of this issue and the state of scientific and technical understanding, we conclude that such analyses are indeed possible within limitations noted in this report. In such cases, these analyses can provide useful guidance for assessing compliance with required health standards, as Chapter 3 of this report will describe.

Even when scientifically useful analysis is possible, assessments of repository performance must contend with substantial uncertainties in information about, and understanding of, the basic physical processes that are important to judging the effectiveness of the repository system to

---

[2] In this report, "biosphere" refers to the region of the earth in which environmental pathways for transfer of radionuclides to living organisms are located and by which radionuclides in air, ground water, and soil can reach humans to be inhaled, ingested, or absorbed through skin. Humans can also be exposed to direct irradiation from radionuclides in the environment.

isolate wastes. Although some of these uncertainties can be resolved by further research, not all of them can be. Some areas — projecting the behavior of human society over very long periods, for example — are beyond the limits of scientific analysis. For these reasons, we have attempted to be candid about the limits of scientific analysis in supporting the standard-setting process. We have made explicit those instances where, because there is no adequate scientific basis for an analysis, policy judgments are required.

Additionally, setting and assessing compliance with a standard must rely on informed judgments and reasonable assumptions based on scientific expertise when uncertainties and unknowns otherwise stand in the way of determinative analysis. There are no alternatives to relying on policy judgments and informed assumptions since some aspects of standard-setting and compliance analysis are not amenable to scientific analysis.

The processes of setting a standard and licensing a repository also raise social, political, and economic issues that would be difficult to resolve even if the scientific challenges were less formidable. Some of these issues might have more effect on the repository program than the health and safety standard itself. Although we have taken a broad view of our charge as related to the scientific basis for the standard, we have not addressed these other, potentially important, issues. The following discussion describes eight issues that we have not addressed.

1. *We have not recommended what levels of risk are acceptable.* A standard that serves as an objective for protection of public health must be stated in terms of some quantitative limit, such as acceptable dose, health effects, or risk. The specific level of acceptable risk cannot be identified by scientific analysis, but must rather be the result of a societal decisionmaking process. Because we have no particular authority or expertise for judging the outcome of a properly constructed social decisionmaking process on acceptable risk, we have not attempted to make recommendations on this important question. However, many domestic and international bodies have reached carefully considered conclusions on this and related questions. We discuss these instances in Chapter 2 and note

the cases where we believe that existing scientific, regulatory, and other expert opinions establish ranges within which lie useful starting points for consistent regulatory proposals.

2. *We have not considered whether the development of a permanent repository should proceed at this time.* A central objective of the DOE program is to license and operate a repository as soon as possible. As individuals, we hold differing views on the urgency of meeting this objective. We were not asked and we did not attempt to address whether a repository is needed in the near future; nor did we compare the risks and benefits of proceeding with a repository now as opposed to those that might be realized by continued reliance on surface storage well into the next century. Accordingly, this report should not be interpreted as a recommendation for or against the development of a Yucca Mountain repository or even a judgment on whether any deep geologic repository should or should not be built at this time.

3. *We have not made a judgment about the suitability of Yucca Mountain as a repository site, or on whether the proposed repository there would meet requirements of any standard consistent with our recommendations to EPA.* Within our scope, we have not produced new scientific or technical data or made calculations that would add to the continuing assessment of the suitability of the site. Although we have reviewed the assessments currently underway, we have not evaluated either the quality or the results of the assessment program in a detailed, rigorous way. Finally, the question of site acceptability raises a variety of social, political, and economic issues that we have not examined because such issues are not within our mandate.

4. *We have not considered the effects of our recommendations on the future of nuclear power.* It has been argued that unless and until means for long-term disposal of spent fuels from commercial nuclear power plants are available, the future of nuclear power is in question. Some states and some foreign countries require by law or regulation that a means

for disposing of waste be in place before additional plants are licensed. We did not, however, consider the effect on the future of nuclear power on the federal program for managing spent fuel from commercial nuclear power plants.

5. *We have not compared the basis for regulating high-level radioactive waste with the basis for regulating nonradioactive long-lived toxic substances, such as lead or cadmium.* Radioactive wastes are sometimes regulated on more stringent bases than nonradioactive wastes even though some nonradioactive substances are more persistent and can pose a greater hazard than many radionuclides. However, it is consistent with our charge in Section 801 to concern ourselves only with the radioactive constituents of the waste.

6. *We have not evaluated the standards applicable to the operational phase of the repository program.* This phase refers to the time before the approved repository is closed and includes the transportation of waste to the repository site and the steps taken at the site to prepare and emplace the waste in the repository. These operations are closely analogous to other nuclear activities regulated by EPA and USNRC. Even though some would argue that the health risk associated with these relatively transitory activities might be greater than those associated with the repository over geologic time, we have not addressed the issues because the clear intent of Section 801 is that our report should focus on the post-operational performance of the repository over very long time-periods. Furthermore, the basis for regulating operating nuclear facilities is considerably better established.

7. *We have not considered the potential effects of the repository on nonhuman biota and ecosystem functions.* These effects might deserve attention, but the clear charge in Section 801 to focus on protection of public health has deterred us from going further. We are aware, of course, and have considered, that human health can be affected by exposure to radionuclides taken up by other organisms such as food crops.

8. *We have not considered the potential for chain reactions of fissile materials as part of a standard.* The possibility

theoretically exists that circumstances might ultimately arise in which radioactive wastes containing fissile materials could undergo a chain reaction in a geologic repository. The potential is an important concern for engineering design that ultimately is likely to be the subject of regulation, perhaps by USNRC. This topic, however, requires specialized analysis that is sufficiently far from our primary focus that we left it for the consideration of others.

BACKGROUND AND APPROACH

A general description of the repository system, and of the ways that it may release radionuclides into the accessible environment, is essential background information for understanding our approach to this assignment. This description appears below, and is followed by discussions of the major issues to be considered in setting a health and safety standard, and of their implications for the study. A map showing the location of the Yucca Mountain region is shown in Figure 1.1. A schematic cross section of the potential Yucca Mountain repository is shown in Figure 1.2.

The Repository System

DOE plans to achieve containment and isolation of high-level radioactive waste in a proposed repository by using an engineered barrier system and locating the repository in the geologic setting of Yucca Mountain. The general repository design suggests that the waste would be emplaced in drifts (tunnels) about 300 meters (1,000 feet) beneath the land surface but above the water table of the uppermost aquifer, that is, in the unsaturated or vadose zone. By law the repository is conceptually designed to hold 70,000 metric tons of high-level radioactive waste. Under current policy, about 90% of this amount (63,000 metric tons) would be spent commercial fuel and the rest would be defense high-level waste. Up to 100 years after emplacement operations begin, the repository would be sealed

Figure 1.1 Map showing location of Yucca Mountain region adjacent to the Nevada Test Site in southern Nevada. Source: Wilson et al., 1994.

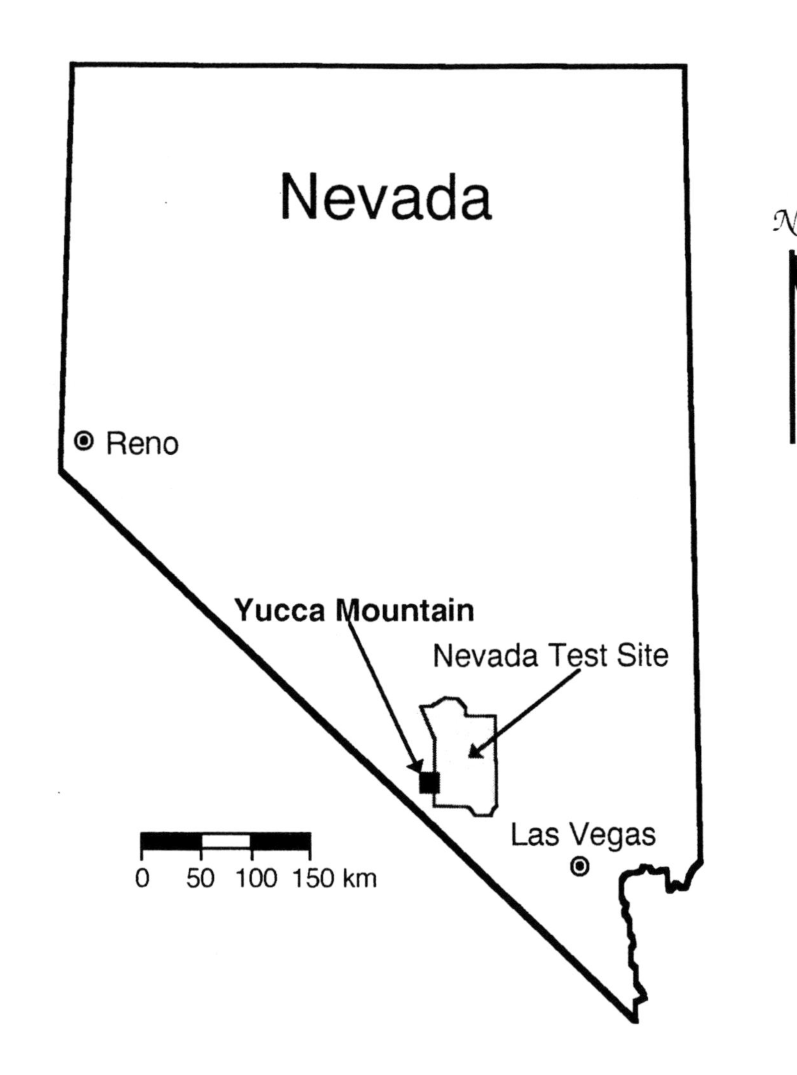

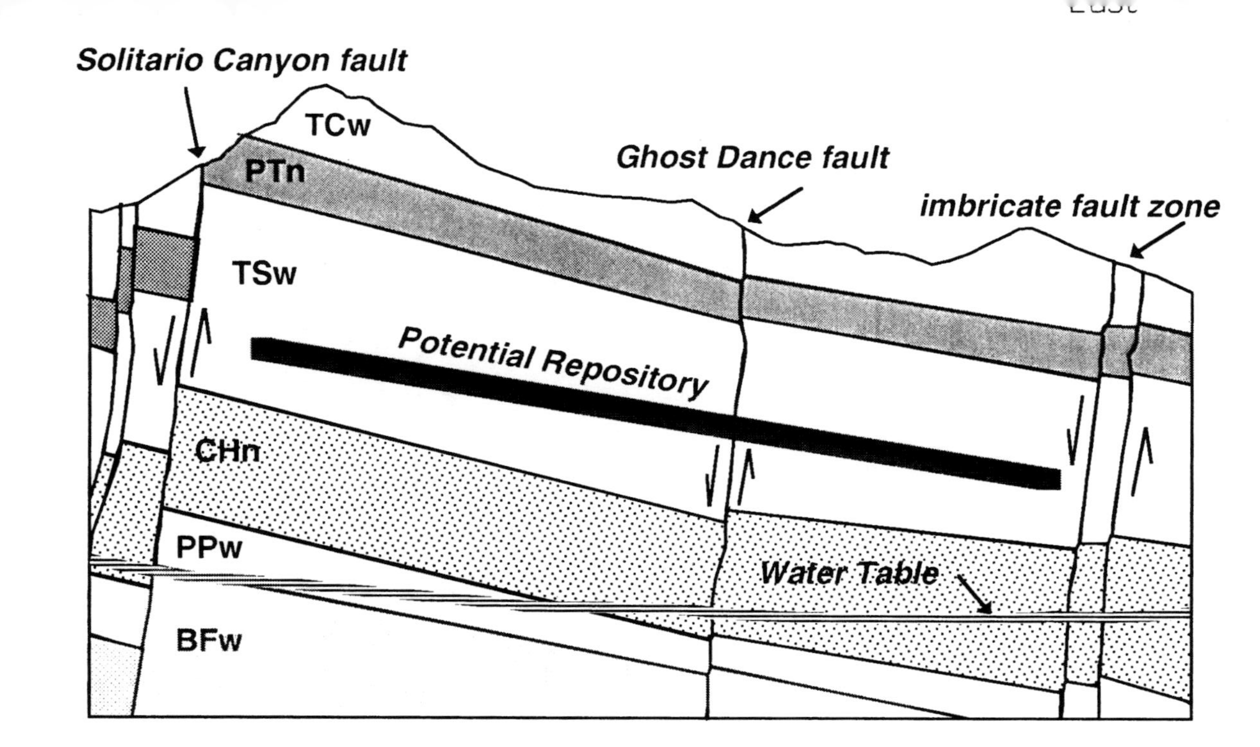

Figure 1.2 Schematic cross section of the potential Yucca Mountain repository region showing location of the repository horizon and static water table with respect to the thermal/mechanical stratigraphic units defined by Ortiz et al. (1985). TCw: Tiva Canyon welded unit; PTn: Paintbrush nonwelded unit; TSw: Topopah Spring welded unit; CHn: Calico Hills nonwelded unit; PPw: Prow Pass welded unit; BFw: Bullfrog welded unit.
Source: Wilson et al., 1994.

by backfilling the drifts, closing the opening to each emplacement drift, and sealing the entrance ramps and shafts.

The engineered barrier system would include the waste form (for example, reactor-fuel assemblies or high-level defense waste embedded in a glass matrix), internal stabilizers, the canister in which the waste is placed, and backfill between the canister and the adjacent host rock. The spent fuel assemblies include naturally radioactive uranium oxide containing fission products, as well as fuel cladding and support hardware, both of which will be radioactive due to activation or contamination. The defense waste consists of products resulting from physical and chemical processes associated with the separation of fissionable materials in weapons manufacture.

The engineered barrier system would be placed beneath Yucca Mountain in the unsaturated zone, which consists of layered units of welded and non-welded tuffs[3]. Some of these units are highly fractured — a characteristic that may influence the flow of water underground. The water table at Yucca Mountain occurs at depths of 600 meters to 800 meters below land surface, which would correspond to depths of 300 to 500 meters below the repository. The volume of rock below the water table contains two principal aquifer systems, one in the volcanic tuff and another at greater depth in carbonate rock. In the Yucca Mountain region, the regional ground water in the upper aquifer appears to flow generally southerly, from higher elevations north of the mountain to the Death Valley region to the southwest where it emerges at the surface (NRC, 1992).

Radionuclide releases from an undisturbed repository into the geologic environs can occur through the following sequence: degradation and failure of the waste canister through corrosion, relatively quick release of substances from the more mobile components of the radionuclide inventory, slow release of substances from the less soluble or less mobile components of the inventory, and movement of radionuclides from the waste package to the air and water in the pores and fissures of the host rock by gas phase and aqueous phase. Radionuclides can enter the environment accessible to humans by traveling down through the unsaturated zone and into the aquifer (the saturated zone), then through the aquifer to wells or springs where the water might be used for purposes such as drinking or agricultural irrigation. Releases might also occur in gaseous form,

---

[3] Tuff is consolidated volcanic ash.

transported upward or laterally from the waste package through the rock to the atmosphere. Other pathways might develop if the site is disturbed, for example, by human intrusion or earthquakes.

More detailed information on the proposed repository and the inventory of radionuclides in the waste is presented in the 1993 total-system performance assessments for Yucca Mountain that were prepared for DOE (Andrews et al., 1994; Wilson et al., 1994).

Issues to Be Considered in Approaching the Study

The aim of this study is to provide guidance on the scientific basis for a standard that would protect the public health from the adverse effects of releases from a proposed repository for high-level radioactive waste at Yucca Mountain. There are two major considerations in providing this guidance. The first is how to make the best use of the scientific knowledge that is now or might soon be available. The second is how to make decisions when the scientific basis is deficient. We present below several examples that illustrate these two considerations, and then describe how we have addressed them in our approach to the study.

Large but improbable doses

It is important to define the standard in such a way that it is a useful measure of the degree to which the public is to be protected from releases from a repository. The nature of geologic disposal is to concentrate and isolate high-level radioactive wastes in a small area for a very long time. It is always possible to conceive of some circumstance that, however unlikely it may be, will result in someone at some time being exposed to an unacceptable radiation dose. Some of these scenarios are common to all geologic repositories; for example, it is always possible that a person will drill or otherwise intrude into any repository in such a way as to bring to the surface some amount of radioactive waste. Other such scenarios are dependent upon the characteristics of the repository site. In the case of Yucca Mountain, human ingestion of radionuclides in ground water drawn from a well is an example of a site-specific scenario that, because of the limited amounts of water in a relatively isolated hydrologic basin, potentially could lead to radiation doses of a relatively high level to a few persons. The possibility that future volcanic activity in the region might seriously compromise the integrity of a repository at Yucca Mountain must also be evaluated. The challenge is to define a standard that specifies a high level of protection but that does not rule out an adequately sited and well-designed repository because of highly improbable events.

Demonstration of compliance

The feasibility of assessing compliance with the standard is another key issue. Quantitative performance assessment is the tool generally proposed for use in evaluating whether a repository is likely to meet the standard with a given level of assurance. Performance assessment requires analyzing the processes by which radionuclides might be released from the repository, the processes by which people might be exposed to them, and the health consequences of exposure. The first steps in the analysis are to model the degradation of waste packages and the migration of radionuclides through the engineered and geologic barriers of the repository and the adjacent host rock. Although this analysis involves important uncertainties, they can, in principle, be addressed by scientific methods. More difficult is the identification of the pathways through the biosphere that would result in exposure to humans. There are countless possible pathways for radionuclides but only a limited number of them need to be analyzed, that is, the ones most likely to yield the highest doses. Moreover, in principle, pathway and exposure analyses require specifying the state of human society many thousands of years into the future — where people might live, what they will eat and drink, what technologies will be available to detect and avoid radionuclides, and other factors. These difficulties cannot be ignored in setting a practical health-based standard, but dealing with them can depend as much, or perhaps more, on assumptions and informed judgment as on testable scientific hypotheses. The scientific basis for performance assessment thus varies considerably among the steps in the analysis.

Fundamental vs. derived standards

To avoid explicitly using uncertain assumptions in compliance assessment, a derived standard is sometimes proposed rather than a fundamental one. A fundamental standard uses as its criterion the endpoint that the standard is intended to control. Thus, when adverse health effects are the outcome to be controlled, a fundamental standard would be stated in terms of limiting the number of adverse effects, the risks of developing an adverse health effect, or of some closely related parameter such as a dose rate. A derived standard translates the fundamental criterion into some other unit of measure, such as the total flux of radionuclides across a repository boundary, expressed for example in the cumulative amount of radioactivity released over a specified period of time.

The difference between the two is that the derived standard subsumes into its definition various assumptions, such as specifying the particular sets of pathways to human exposure, and a dose-response relationship, that would otherwise have to be made in compliance assessment for a fundamental standard. Because a derived standard might eliminate from the licensing process some of the calculations involved in specifying these pathways, it has the advantage of a simpler licensing decision (M. Federline, USNRC, personal communication, May 27, 1993). In choosing between a fundamental or a derived standard, a balance must be struck between clarity of purpose in the standard and complexity of the licensing process on the one hand, and complexity in the standard, but a clearer focus in the licensing process on the other.

Time scale

A final issue involves the time scale over which compliance with the standard should apply. The repository could release radionuclides over hundreds of thousands of years or more, but as performance assessments are extended into the future, the uncertainties in some of the calculations that might be required could render further calculation scientifically meaningless. On the other hand, analyses that are uncertain at one time might not be so uncertain at a later time; for example, the uncertainties about cumulative releases to the biosphere that depend on the rate of failure of the waste packages are large in the near term but are smaller later, when enough time has passed that all of the packages will have failed. Selection of a time scale for the standard must therefore take into account the scientific basis for the performance assessment itself. Selection of a time scale also involves policy considerations. (For example, the level of protection that the standard affords to future generations is an important ethical question that must be considered. Limiting the time period covered by the standard could be inconsistent with a policy on long-term intergenerational equity.)

The remanded EPA standard — and the recently promulgated standard for radioactive waste repositories other than the proposed Yucca Mountain repository — places a time limit on performance assessment of 10,000 years. This time limit makes some aspects of the analysis more tractable by eliminating from consideration the uncertainties that increase at times beyond 10,000 years. In the case of Yucca Mountain, however, recent performance assessment calculations (Andrews et al., 1994) indicate that the likely time for some radionuclides, such as technetium-99, to reach the biosphere is longer than 10,000 years. If that time limit were to apply at the Yucca Mountain site, potential exposures occurring beyond 10,000 years would be excluded from the compliance analysis. The problem of the cumulative uncertainties must therefore be weighed against the need to consider the exposures when they actually are calculated to occur.

Choices Affecting the Bases of the Standard

The foregoing issues illustrate two considerations that we have had to balance in reaching our conclusions and recommendations. First, is the need to choose among the available options (for example, alternative forms of the standard and time scales) in a way that makes the best use of the scientific information that is available. For example, it might be intuitively attractive to state a standard in terms of risk to human health. But as noted earlier, the demonstration of compliance with such a standard requires a model of the radionuclides and their pathways from the repository to the biosphere that is scientifically challenging to develop. This difficulty can be avoided by abandoning a health-based standard in favor of a limitation on releases from the repository, but doing so would obscure crucial information about the potential of the radionuclide releases for causing health effects. Similarly, selecting a time scale for analysis involves weighing how the scientific basis for analysis changes with time against the timing at which more numerous future health effects are likely to occur. We have tried to deal explicitly with these choices and to arrive at a basis for judging the form of standard that is best supported by the available scientific information taken as a whole.

The second consideration is how to provide, within the regulatory process, a system for making those choices for which scientific information is unavailable or insufficient. The regulatory process involves the two major steps of rulemaking and licensing. The rulemaking procedure allows extensive public participation and considerable administrative discretion in weighing and assimilating alternative points of view. Licensing is a quasi-judicial process that benefits from having clear-cut limits against which to judge an applicant's proposals. It is for the latter reason that several members of the USNRC staff have pointed out their reluctance to leave any speculation about the future of human society for the licensing process (which USNRC administers).

There are several choices to be made in designing the standard for which science cannot provide all the necessary guidance — defining the critical group to be protected or the radionuclide pathways to them through the biosphere, for example. Since these choices must be made, even in the absence of clear-cut scientific information, we recommend that such issues should be treated as part of the rulemaking process, since this process, as indicated earlier, allows a broader scope for discussing and weighing alternatives.

In the course of this study, we analyzed separately the scientific bases for setting a health-based standard, conducting compliance assessment, and dealing with human intrusion and episodic geologic processes, such as volcanoes and earthquakes. We adopted this procedure to help us understand the choices involved among these different aspects of the problem, and to clarify where the scientific basis for choice was insufficient. We then weighed these considerations in making our final findings and recommendations, which are presented in the remaining chapters of our report.

2

PROTECTING HUMAN HEALTH

The primary objective of the proposed repository at Yucca Mountain is to dispose of high-level radioactive defense waste and spent nuclear fuel in a safe manner. To determine whether the repository can be designed to protect the public health from the risks associated with exposure to radiation from radionuclides that may be released from the repository, it is necessary to establish standards against which to judge whether the design of the repository is acceptable. This target will be embodied in a radiation protection standard to be issued by EPA.

In Section 801 of the Energy Policy Act of 1992, Congress directs that EPA set these standards by specifying the maximum annual effective dose equivalent to individual members of the public. In the same section, Congress also asks three questions, the first of which is:

> whether a health-based standard based on doses to individual members of the public from radionuclide releases to the accessible environment . . . will provide a reasonable standard for the protection of the health and safety of the general public.

This chapter addresses this question. As background, we first present a synopsis of the health effects of ionizing radiation and outline the development of radiation protection standards on a national and international basis. This discussion will illustrate the current status of scientific investigation and consensus of expert judgment on which most efforts to establish a standard for high-level waste repositories are based.

We then turn to the question of whether a standard for Yucca Mountain designed to protect individuals will, if met, also protect the general public. We conclude that the answer to this question is "yes," given the particular characteristics of the site and assuming that policy makers and the public are prepared to accept that very low radiation doses pose a negligible risk.

Because the current EPA standard for nuclear waste disposal in 40 CFR 191 takes an approach different from that required by Congress, however, addressing only the question posed in Section 801 is too narrow a response. Accordingly, we have expanded the discussion by recommending the use of a standard designed to limit individual risk rather than individual dose and by describing how a standard might be structured on this basis. We then address the specific question of protection of public health in the context of an individual-risk standard and compare this standard with the one currently used by EPA for sites other than Yucca Mountain. Based on this analysis, we conclude not only that an individual-risk standard would protect the health of the general public, but also that this form of standard is particularly appropriate for the Yucca Mountain site in light of the site's characteristics.

Finally, standards are only useful if it is possible to make meaningful assessments of future repository performance with which the standards can be compared. In Chapter 3, we discuss our conclusion that it is feasible to conduct such compliance assessments against an individual-risk standard. Doing so, however, requires using the rulemaking process to arrive at a regulatory decision about certain assumptions as part of the standard, for example., about future human behavior. In the following discussion of the standard, we have indicated the assumptions for which this is required.

THE HEALTH EFFECTS OF IONIZING RADIATION

Cell and gene damage can be caused in humans exposed to ionizing radiation (NRC, 1990a), (also referred to as the BEIR V report). Extremely high doses of radiation can lead to quick death, as seen, for example, in Nagasaki, Hiroshima, and Chernobyl. However, even much lower levels of radiation can affect health. International scientific bodies currently accept what is called the linear, or no-threshold hypothesis for the dose-response relationship. Most of what is known about effects of radiation on human health comes from studying people exposed to large doses of radiation. The empirical relationship between cancer induction and radiation dose appears linear at the high doses received by the atomic bomb survivors. The linear hypothesis postulates that this dose-response relationship continues when extrapolated to very low doses. The no-threshold hypothesis holds that there is no dose, no matter how small, that does not have the potential for causing health effects. To explain this relationship of radiation to cancer, and other health effects, the following outlines the interaction between radiation and the human body.

Radiation that is sufficiently energetic to dislodge electrons from an atom is referred to as ionizing radiation. Impinging ionizing radiation, colliding with atoms and molecules in its path, gives rise to ions and free radicals that break chemical bonds and cause other molecular alterations in affected cells. Any molecule in the cell can be altered by radiation, but deoxyribonucleic acid (DNA), the double helix of base pairs that make up the genes to be passed on to the next generation, is the most critical molecular target because of the uniquely important genetic information it contains. Damage to a single gene, which might consist of thousands of base pairs, can profoundly alter or kill the cell. Although millions of changes in DNA are produced in the body of every person each year by exposure to natural background radiation and other influences, most of the changes are reparable. If unrepaired or misrepaired, however, the damage might be expressed in the form of permanent genetic changes or mutations, the frequency of which approximates 10^{-5} to 10^{6} per gene per Sievert (Sv)[1]. Because the mutation rate tends to change in direct proportion to the dose, it is inferred that the interaction of the gene with a single ionizing particle might suffice in principle to mutate the gene. Damage to the genetic apparatus of a cell can also cause changes in the number or structure of its chromosomes, the thread-like structures on which the genes are arranged. Such changes increase in frequency in proportion to the dose in the range below 1 Sv.

Radiation damage to genes, chromosomes, or other vital organelles can be lethal to affected cells, especially dividing cells, which are highly radiosensitive as a class. The survival of dividing cells, measured in terms of their capacity to grow and divide, tends to decrease exponentially with increasing dose, 1-2 Sv generally sufficing to reduce the surviving cell population by about 50% (NRC, 1990a). The killing of cells, if sufficiently extensive, can impair the function of the affected organ or tissue. In general, however, too few cells are killed by a dose below 0.5 Sv to cause clinically detectable impairment of function in most human organs other than those of the embryo. Because such effects on organ function are not produced unless the radiation dose exceeds an appreciable threshold, they

---

1 A unit of equivalent radiation dose, a Sievert is the product of the absorbed dose and the radiation weighting factor. 1 Sievert equals 100 rem.

are commonly viewed as nonstochastic (or deterministic) effects, in contradistinction to mutagenic effects of radiation, which are viewed as stochastic effects because they might have no thresholds (see Glossary). Carcinogenic effects of radiation, which can result from mutational changes in the affected cells, are likewise viewed as stochastic effects, the frequency of which is assumed to increase as a linear, no-threshold function of the dose, although the possible existence of a threshold for such effects cannot be excluded.

Natural background radiation is estimated by the National Council on Radiation Protection and Measurements (NCRP) to contribute 82% of the average annual radiation exposure to a United States citizen, and medical applications, an additional 15% (NCRP, 1987a). All other sources of radiation exposure together contribute approximately 3% (Table 2-1). All sources combined give an average dose of 3.6 mSv/yr (360 mrem/yr). Background radiation levels are not uniform. For example, the average difference in background radiation between Denver, CO and Washington, DC, is 0.3 mSv/yr (30 mrem/yr). One cross-country plane ride contributes approximately 0.025 mSv (2.5 mrem) (NCRP, 1987a,b).

At the low-dose rates characteristic of natural background radiation or occupational irradiation, the only health effects of radiation to be expected are stochastic effects; that is, mutagenic and carcinogenic effects. Although the risks of certain cancers have been significantly elevated in some cohorts of radiation workers, especially those employed in the era preceding modern safety standards, no definite or consistent evidence of carcinogenic effects has been observed in workers exposed within present maximum permissible dose limits or in populations residing in areas of high natural background radiation. Hence, assessment of any cancer risks attributable to irradiation in such populations must be based on extrapolation from observations of the effects of exposure at higher dose levels. Because a statistically significant increase in heritable abnormalities is yet to be demonstrated in human beings at any dose level, assessment of the risks of such effects must be based on extrapolation from observations on laboratory animals. Because of the assumptions inherent in the extrapolations that are involved, assessments of the carcinogenic and mutagenic effects of low-level irradiation are highly uncertain. The uncertainties notwithstanding, it has been possible to reach a reasonable consensus within the scientific community on the relationship between doses and health effects, that is generally considered to provide an acceptable basis for evaluating the risks attributable to a given dose or the degree of protection afforded by a given limitation of exposure.

Within recent years, the risks attributable to low-level irradiation have been assessed in detail by the United Nations Scientific Committee on the Effects of Atomic Radiation (UNSCEAR, 1988), the National Research Council Committee on the Biological Effects of Ionizing Radiation (NRC, 1990a), and the International Commission on Radiological Protection (ICRP, 1991). The last of these assessments, which drew on and extended the previous two, arrived at risk assessments for carcinogenic effects and for heritable effects, which are shown in Table 2-2. Carcinogenic effects, which are expressed only in exposed individuals themselves, are estimated to account for the bulk (80%) of the overall risk of harm. The lifetime risk of developing a fatal cancer from irradiation is estimated to be 5×10^{-2}/Sv for a member of the general population. Nonfatal cancers, although projected to be produced more frequently than fatal cancers, were judged to contribute less to the overall health impact of irradiation because of their lesser severity in affected individuals and were, therefore, weighted accordingly (Table 2-2). Of the total risk of heritable effects, about one-fourth is projected to be expressed in the first two generations alone, the remainder during subsequent scores of generations.

This table indicates that if 100 people were each to receive 1 Sv of radiation over their lifetimes, which is about 300 times greater than the overall average annual natural background level of radiation in the United States, five would be expected to die from cancer induced by that radiation. Since it accounts for the great bulk of the potential harm that might be attributed to low-level radiation, the above risk estimate for fatal cancer is often used to calculate the expected number of fatalities attributable to low-dose irradiation in a population. For example, if one million persons were each exposed to a dose equivalent to that received from a transcontinental plane ride (0.025 mSv), the resulting collective dose (25 person-Sv) would be estimated to cause one extra fatal cancer in the population in addition to the 200,000 fatal cancers that would be expected to occur in the same population from all other causes combined. Because the added risk, if any, is calculated to be such a small fraction of the total cancer risk, it is not surprising that epidemiological data have revealed no significant differences in the rates of cancer or other diseases among populations exposed to far larger variations in natural background radiation levels (NRC 1990a).

Table 2-1 Average Amounts of Ionizing Radiation Received Yearly by a Member of the U.S. Population[a]

Source	Dose[b]	
	(mSv/yr)	(%)
Natural		
Radon[c]	2.0	55
Cosmic	0.27	8
Terrestrial	0.28	8
Internal	0.39	11
Total Natural	3.0	82
Anthropogenic		
Medical		
X-ray diagnosis	0.39	11
Nuclear medicine	0.14	4
Consumer products	0.10	3
Occupational	< 0.01	< 0.3
Nuclear fuel cycle	< 0.01	< 0.03
Nuclear fallout	< 0.01	< 0.03
Miscellaneous[d]	< 0.01	< 0.03
Total anthropogenic	0.63	18
Total Natural and Anthropogenic	3.6	100

[a] From NRC (1990a) and NCRP (1987a)
[b] Average effective dose equivalent
[c] Dose to bronchial epithelium alone
[d] DOE facilities, smelters, transportation, etc.

Table 2-2. Estimated Frequencies of Radiation-Induced Fatal Cancers, Nonfatal Cancers, and Severe Hereditary Disorders, Weighted for the Severity of their Impacts on Affected Individuals[a]

	No. of cases per 100 per Sv[b]
Fatal cancers	5.0
Nonfatal cancers	1.0
Severe heredity disorders	1.3
Total	7.3

[a] From ICRP (1991)
[b] Numbers of cases, weighted for severity of their impacts on affected individuals over their lifetimes, attributable to low-level irradiation of a population of all ages.

DEVELOPMENT OF RADIATION PROTECTION STANDARDS

There is a worldwide interest in the development of radiation protection standards, including those for the disposal of high-level radioactive waste, and a considerable body of analysis and informed judgment exists from which to draw in formulating a standard for the proposed Yucca Mountain repository. EPA's process for setting the Yucca Mountain standard is presumably not bound by this experience, but a sound technical approach should include a review of other relevant work to date. Accordingly, we summarize below the status of relevant work on radiation protection standards both in the United States and abroad.

General Consensus in Radiation Protection Principles and Standards

A number of international and nongovernmental national bodies (such as the International Atomic Energy Agency (IAEA), ICRP and NCRP) have recommended radiation protection principles and standards.

These recommendations, in turn, usually are considered by the national agencies that set radiation protection standards, which then are codified into pertinent rules and regulations. Of the international bodies, the International Commission on Radiological Protection (ICRP) is perhaps the most influential. Its counterpart in the U.S. is the National Council on Radiation Protection and Measurements (NCRP).

In the United States, several agencies establish radiation protection standards in their areas of responsibility. Among them are the following: the U.S. Environmental Protection Agency (EPA), the U.S. Nuclear Regulatory Commission (USNRC), and the U.S. Department of Energy (DOE). These three agencies play key roles in programs involving public health and safety, environmental protection, health and safety in the nuclear industry, and radioactive waste management and disposal.

Recommendations for radiation standards to protect the public health and safety are frequently based on the analyses of radiation risks developed by the United Nations Scientific Committee on the Effects of Atomic Radiation (UNSCEAR) and the ICRP on the international level and by the Committees on Biological Effects of Ionizing Radiation (BEIR) in the United States. The most recent analyses are presented in the UNSCEAR (1988) and NRC (1990a) reports, respectively.

Concurrent with the development of radiation protection concepts internationally and in this country, a consensus has emerged among the organizations involved in performing analyses and making recommendations (ICRP, NCRP, NRC's BEIR V, and UNSCEAR) and those that promulgate regulations (EPA, USNRC, and DOE). This coalescence of views and resulting consensus can be seen in the general uniformity in the system of radiation dose limitation, fundamental units and terminology, health effects factors, occupational and public dose limits, dose apportionment, and use of the critical-group concept. The latter two concepts are defined and discussed later in this chapter.

Consistent with the current understanding of the related consequences, ICRP, NCRP, IAEA, UNSCEAR, and others have recommended that radiation doses above background levels to members of the public not exceed 1 mSv/yr (100 mrem/yr) effective dose for continuous or frequent exposure from radiation sources other than medical exposures. Countries that have considered national radiation protection standards in this area have endorsed the ICRP recommendation of 1 mSv per year radiation dose limit above natural background radiation for members of the public. In the United States, DOE, in Order 5400.5, and USNRC, in 10 CFR 20, have set the dose standard for public exposure to ionizing radiation at 1 mSv per year above natural background level. EPA is in the process of developing similar guidance for all U.S. federal agencies (EPA, 1993).

This framework, with an effective dose limit of 1 mSv per year, is used as a basis for protecting the public health from routine or expected anthropogenic sources of ionizing radiation (i.e., resulting from human activity) other than medical exposures. It includes any exposures to the public derived from the management and storage of high-level radioactive defense waste and spent nuclear fuel. We note that guidance to date has been for expected exposures from actual routine practices. There is little guidance on potential exposures in the far distant future.

ICRP (1985a) proposed apportionment of the total allowable radiation dose from all anthropogenic sources of radiation, excluding medical exposures. Thus, for radioactive waste management, including high-level radioactive defense waste and spent nuclear fuel, the national authorities could apportion, or allocate, a fraction of the 1 mSv per year to establish an exposure limit for high-level waste facilities. EPA in 40 CFR 191 noted that its requirement for the WIPP transuranic waste facility, at a level of 0.15 mSv/yr (15 mrem/yr), is consistent with ICRP's concept of apportionment.

Most other countries also have endorsed the principle of apportionment of the total allowed radiation dose. Apportionment values that have been established by various countries for high-level radioactive waste range from 5% to 30%, corresponding to radiation doses ranging from 0.05 mSv (5 mrem) per year to 0.3 mSv (30 mrem) per year.

Table 2-3 presents the limits established by various countries on individual exposure from high-level waste disposal facilities. The information in this table suggests a general consensus among national authorities and agencies to accept and use the principle of radiation dose apportionment.

THE FORM OF THE STANDARD

A standard is a societally acceptable limit on some aspect of repository performance that should not be exceeded if the repository is to be judged safe. There is, however, a variety of ways in which this limit can be formulated. It can, for example, be imposed at several points in the chain of events that might ultimately lead to adverse effects on public health. Thus, the limit could apply to the amount of radionuclides released from the repository, to the radiation doses to persons resulting from those releases, to the number of health effects associated with the doses, or to the level of risk. Risk, dose, or health effect limits can be stated for individuals or for populations.

We recommend the use of a standard that sets a limit on the risk to individuals of adverse health effects from releases from the repository. In this context, risk is the probability of an individual receiving an adverse health effect. It is essential to define specifically how to calculate this risk, however, for otherwise it will not be clear what number to use to compare with the risk limit established in the standards.

From the scientific perspective, the calculation of health risks should take into account all of the uncertainties involved in analyzing repository performance over very long time periods. Because many of the elements of the calculation are not well known, they must be dealt with by using distributions that represent the analysts's state-of-knowledge. The first step in calculating risk is therefore to develop a distribution of doses received by individuals, taking into account all of the events that go into determining whether a dose is received.[2] A probabilistic distribution of the health effects associated with these doses can then be developed as the product of each value of dose received and the health effects per unit dose. In this report, we choose to define risk as the expected value of the probabilistic distribution of health effects.[3]

---

2 This does not mean that every event needs to be treated probabilistically; some might be represented by a single bounding estimate, for example. The definition does require, however, that all of the parameters that determine the dose be considered in developing the probabilistic distribution of dose.

3 It is both easier and common practice to calculate doses received over an individual lifetime. One reason is that the effects of radiation might not appear until years after the dose is received. The lifetime calculation can be annualized by dividing by the duration of an average lifetime. Since this annualized risk is often more convenient for comparison to other risks, we recommend it be used.

Table 2-3 Quantitative High-Level Waste Disposal Objectives/Criteria at International Level and in OECD Countries[a]

Organization/Country	Main Objective/Criteria	Other Main Feature(s)	Comments
NEA (1984)	Max. indiv. risk objective 10^{-5}/yr (all sources)	Individual risk/dose = best criterion to judge long-term acceptability	No consensus on ALARA/optimization
ICRP Publication 46 (1985)	1 mSv/yr (normal evolution scenarios) 10^{-5}/yr (probabilistic scenarios) for individuals (all sources)	Both prob. and doses should be taken into account in ALARA	ALARA useful, notably to compare alternatives, but might not be the most important siting factor
IAEA Safety Series 99 (1989)	ICRP Publication 46		Also includes qualitative technical criteria on disposal system features and role of safety analysis and quality assurance
CANADA AECB regul. Document R.104 (1987)	Max. indiv. risk obj. 10^{-6}/yr	Period of time for demonstrating 10^4yr No sudden and dramatic increase for times > 10^4yr	Additional qualitative, nonprescriptive requirement and guidelines in regulatory documents No explicit optimization required

FRANCE	Under development[b]: Ref. to ICRP Publication 46		Technical criteria for siting established in 1987
GERMANY Section 45, para 1 of Radiation Protection Ordinance (1989)	Individual dose < 0.3 mSv/yr for all reasonable scenarios	Calculation of individual doses limited to 10^4yr but isolation potential beyond 10^4yr might be assessed	Additional qualitative technical criteria in guidelines and regulatory documents
THE NORDIC COUNTRIES Consultative document (1989)	Individual dose < 0.1 mSv/yr (normal scenarios) Individual risk < 10^{-6}/yr (disruptive events)	Additional criterion on "total activity inflow" limiting releases to biosphere, based on inflow of natural alpha radionuclides	Under revision following broad consultation Includes other qualitative criteria
SPAIN Statement by Nuclear Safety Council, 1987	Individual dose <0.1 mSv/yr Individual risk <10^{-6}/yr in any situation		Further development under study
SWITZERLAND Regulatory Document R-21 (1980)	Individual dose <0.1 mSv/yr at any time for reasonably probable scenarios; individual risk < 10^{-6}/yr for sources with lower probability	Repository must be designed in such a way that it can at any time be sealed within a few years without the need for institutional control (for all times)	Flexibility to amend dose and risk limits depending on numbers exposed

UNITED KINGDOM[c]	No specific criteria for HLW but likely application of principles similar to existing objectives for L/ILW:< 10^{-6}/yr target for individual risk from a single facility	No time-frame for quantitative assessment specified	ALARA to be used to the extent practical and reasonable
U.S. EPA 40 CFR 191 (1985)	Limits on projected radionuclides releases to the accessible environment for 10^4yr, based on objective to limit serious health effects to less than 10 in the first, 10^4yr after disposal for each 1,000 metric tons of heavy metal or other unit of waste	Individual dose (over 1000 yr)$\leq$0.25 mSv/yr Other requirements on drinking water contamination	1985 EPA standard was vacated in 1987 and most of its provisions adopted into law in 1992

U.S. EPA 40 CFR 191 (1993)	Same as in 1985 standard	10,000 year period Individual dose from all environmental pathways $\leq$ 0.15 mSv/yr Requirements to protect underground sources of drinking water to the maximum contaminant level	Not applicable to Yucca Mountain Same as 1985 standard except for individual dose and ground water provisions
U.S. NRC 10 CFR 60	Minimum levels of performance: Waste package ("substantially complete" containment for 300- 1000y) Engineered barrier system (releases $<10^{-5}$/yr of the inventory at 1000 yr after repository closure) Pre-waste-emplacement ground water travel time between "disturbed zone" and "accessible" environment >100y		NRC subsystem requirements are intended to help achieve compliance with the EPA standard and alternative criteria may be approved if appropriate

[a] This table was established by the OECD/Nuclear Energy Agency (NEA) Secretariat, based on national presentations made at a Joint Radioactive Waste Management Committee and Committee on Radiation Protection and Public Health

Workshop on Radiation Protection and Safety Criteria for the Disposal of High-Level Waste, Paris, Nov. 5-7, 1990. It presents national criteria in a very simplified form, and should always be read in conjunction with the descriptions reproduced in the Workshop Proceedings published by NEA. Despite apparent differences, all criteria share the same common basis and aim at a relatively uniform safety level.

[b] France has since adopted a limit of 0.25 mSv/yr. (Dejonghe, 1993).

[c] The UK National Radiological Protection Board has made recommendations for changes in this regard (Barraclough, 1992). As of June 1, 1995, these recommendations are under considerations by the government.

To illustrate, current scientific understanding indicates that the lifetime risk of developing a fatal cancer (based on the dose-response relationship shown in Table 2-2) is $5x10^{-2}$ per Sv. Thus, if the expected value of the lifetime dose that an individual receives, calculated from a probabilistic distribution of dose, is $1x10^{-4}$ Sv, then that person's lifetime risk of a fatal cancer is $5x10^{-6}$ ($1x10^{-4}$ Sv x $5x10^{-2}$ fatal cancers per Sv).

We recognize that our recommendation to use an individual-risk standard differs from the form of standard set by EPA in 40 CFR 191 and is a refinement of the form of the effective-dose standard required by Section 801. At the end of this chapter, we discuss our reasons for preferring the risk-based approach.

ELEMENTS OF AN INDIVIDUAL-RISK STANDARD

We now turn to a discussion of how the key elements of an individual-risk standard for Yucca Mountain might be structured. In particular, it is necessary to specify what level of protection is to be afforded, who is to be protected, and for how long. Establishing this structure is prerequisite to assessing whether the individual-risk standard will protect the health of the general public.

As background for this discussion, it is useful to review some of the relevant characteristics of the Yucca Mountain site. The proposed repository would be located in volcanic tuff several hundred meters above the local water table. When materials are released from the waste packages in the repository, they will be transported downward through an unsaturated zone toward the underlying aquifer by water that infiltrates from the surface. The amount of infiltration or recharge depends on climatic conditions. In the absence of fast transit pathways such as faults, fractures, or drill holes, current understanding suggests that transit times to the water table will be long, perhaps 10,000 to 100,000 years (DOE, 1988).

Once radionuclides reach the aquifer, they would be transported away from the vicinity of the repository in the direction of ground-water flow, which is generally to the southwest from the site. Thus, within the aquifer, there would be a plume of contaminated ground water stretching away from the vicinity of the repository. Near Yucca Mountain, there is no flowing surface water that might serve as a source in preference to ground water. From what currently is known about the aquifer and its low recharge rate, it seems likely that at some times in the future the concentrations of radionuclides in this plume could be relatively high compared with concentrations that would result if the ground water were discharged into a body of flowing surface water (NRC, 1983).[4]

According to current understanding, there are three potential routes by which radionuclides in the ground-water plume could expose humans to radiation. One is through withdrawal of contaminated ground water via wells for local use. Another is through contact where the ground water eventually emerges at the surface. A third would occur if ground water were withdrawn and transported away from the region for use elsewhere. In the judgment of most analysts to date, the most probable route for exposing humans to radiation via ground water at Yucca Mountain is via wells.

In addition to exposure via ground water, humans could also be exposed as a result of gaseous emissions from the Yucca Mountain site. Because the proposed repository is above the local water table, some carbon-14 (^{14}C), the radioactive isotope of carbon, will be emitted as

---

[4] The concentrations of radionuclides in undiluted ground water are likely to be high in the vicinity of almost any repository at some times in the future. A distinguishing feature of the Yucca Mountain site is that there are no surface water sources that would dilute the concentrations of radionuclides if the ground water were discharged to them.

gaseous carbon dioxide, which can migrate through the overlying rock to the surface. Once in the atmosphere, the radioactive carbon dioxide will eventually be distributed across the globe in times relatively short compared with the half-life of ^{14}C. Current understanding suggests that the major pathway for exposure of ^{14}C to humans is through food crops.

What Level of Protection?

The level of protection established by a standard is a statement of the level of risk that is acceptable to society. We acknowledge that determining what is acceptable is not ultimately a question of science but of public policy. Whether posed as "How safe is safe enough" or as "What is an acceptable level of risk?", the question is not solvable by science alone. The rulemaking process, directly involving public comment to which an agency must respond, is an appropriate method of addressing the question of an appropriate level of protection. Accordingly, we do not directly recommend a level of acceptable risk. We do, however, describe the spectrum of regulations already promulgated that imply a level of risk, all of which are consistent with recommendations from authoritative radiation protection bodies.

For example, EPA has already used a risk level of $5x10^{-4}$ health effects in an average lifetime, or a little less than 10^{-5} effects per year, assuming an average lifetime of 70 years, as an acceptable risk limit in its recently published 40 CFR 191. This limit is consistent with other limits established by other U.S. nuclear regulations, as shown in Table 2-4. In addition, the risk equivalent of the dose limits set by authorities outside the United States (shown in Table 2-3) is also in the range of 10^{-5} to 10^{-6}/yr (except for exposure to radon indoors or releases from mill tailings). This range could therefore be used as a reasonable starting point in EPA's rulemaking.

Who Is Protected?

To determine whether a repository complies with the standard, it is necessary to calculate the risk to some individual or group of individuals and then to compare that number with the risk limit established in the

Table 2-4 Comparison of the Annual Individual Risks Associated with USNRC and EPA Standards
Adapted from: Kitty Dragonette, USNRC, personal communication, June 16, 1993 and 40 CFR 191.

Standard	Limit	Annualized Individual Risk[a]
Indoor Radon	4 pCi/1 (0.1 Bq/l)	$4x10^{-4}$
40 CFR 192 (Mill Tails)	20 pCi/M^2s (0.7 Bq/m^2s) 5 pCi $^{226}Ra/g$ (0.2 Bq/g)	$1x10^{-6}$ $3x10^{-4}$
10 CFR 61 (Low Level Waste)[b]	25 mrem/yr (0.25 mSv/yr)	$1.25x10^{-5}$
40 CFR 190 (Uranium Fuel Cycle)	25 mrem/yr (0.25 mSv/yr)	$1x10^{-5}$
40 CFR 191.03 (Repository Operations)	25 mrem/yr (0.25 mSv/yr)	$1x10^{-5}$
40 CFR 191.15 (High-Level Waste Individual Protection Standards)	1985: 25 mrem/yr (0.25 mSv/yr)	$1x10^{-5}$
	1993: 15 mrem/yr (0.15 mSv/yr)	$7.5x10^{-6}$
40 CFR 61 (National Emission Standards for Hazardous Air Pollutants)	10 mrem/yr (0.1 mSv/yr)	$5x10^{-6}$
40 CFR 191.16 (Ground Water Protection Standards)[c]	1985: 4 mrem/yr (0.04 mSv/yr)	$2x10^{-6}$
	1993: Safe Drinking Water Act	$2x10^{-6}$
40 CFR 300 (Superfund)	General 5 pCi $^{226}Ra/g$ (0.2 Bq/g)[d]	10^{-6} to 10^{-8}[e] $3 x 10^{-4}$

[a] Assumes a lifetime risk of $5x10^{-2}$ per Sievert ($5x10^{-4}$ per rem). With two exceptions, the risks in this table are those allowed for an assumed maximally exposed individual. One exception is the reactor safety goal, which is based on average risks experienced by the

population potentially affected by the facility. Translation from average to maximum individual risks (or vice versa) is not possible without specific demographic information about the exposed population. Another exception is 40 CFR 191.13, which is based on collective-dose considerations.

[b] Neglects consideration of ALARA radiation protection measures; actual doses to members of the public from all pathways are generally far below the dose limit.

[c] These levels of the standard are consistent with EPA's ground water protection strategy.

[d] The Superfund requirements address the risk of fatal and nonfatal cancer over a lifetime. In order to present risk values on a consistent basis in the table, the risk is expressed in terms of fatal cancers per year assuming a 70-year lifetime and a ratio of 1.5 for total cancer incidence to fatal cancer incidence. Depending upon exposure pathways, radionuclide, total inventory, and site characteristics, the ratio of 1.5 could be off by a factor of 2.

[e] As applied at selected Superfund sites with ^{226}Ra contamination, for example, Montclair, NJ, Denver, CO.

standard. Therefore, the standard must specify the individual or individuals for whom the risk calculations are to be made. The issue is how to define who is to be protected among the persons having the highest risk of health effects due to releases from a repository, since by definition all other persons face a lower risk.

The choice of those to be protected can obviously have a significant effect on the calculated risk and, therefore, on whether the calculated performance meets the standard. For example, some groups of persons are particularly sensitive to exposure due to factors such as pregnancy, age, or existing health problems. Similarly, it is possible to construct scenarios in which an individual could receive a very high dose of radiation, even though only one or two people might ever receive such doses.

There is an obviously sensitive issue involved here, since the definition of the person or persons to be protected directly affects the outcome of the risk calculation. Although not a purely scientific issue, we believe that a reasonable and practicable objective is to protect the vast majority of members of the public while also ensuring that the decision on the acceptability of a repository is not prejudiced by the risks imposed on a very small number of individuals with unusual habits or sensitivities. The situation to be avoided, therefore, is an extreme case defined by unreasonable assumptions regarding the factors affecting dose and risk, while meeting the objectives of protecting the vast majority of the public. An approach consistent with this objective that is used extensively elsewhere in the world is to define and protect a critical group; we recommend this approach for the Yucca Mountain standards.

The critical group has been defined by the ICRP (1977, 1985b) as a relatively homogeneous group of people whose location and habits are such that they are representative of those individuals expected to receive the highest doses[5] as a result of the discharges of radionuclides. Therefore, as the ICRP notes, "because the actual doses in the entire population will constitute a distribution for which the critical group represents the extreme, this procedure is intended to ensure that no individual doses are unacceptably high." (ICRP 1985a, at paragraph 46). In the case of Yucca Mountain, these individuals presumably would live in the near vicinity of the site and would potentially be exposed to radiation through the use of contaminated ground water.

The critical-group dose is defined as that dose received by an average member of the critical group. Using the average member of the group as the basis for comparison with the limit established by the standard avoids the problem of the outcome being unduly influenced by the habits of a few persons. To ensure that this calculation is nevertheless representative of the persons who receive the highest doses, the ICRP definition of the critical group requires that:

1. The persons calculated to receive the highest doses based on cautious, but reasonable, assumptions be included in the group.
2. The group be homogeneous in dose; that is, there should be a relatively small difference between those receiving the highest and lowest doses in the group (ICRP, 1991). In its Publication 43, the ICRP (1985b) suggests that if the ratio of the calculated average critical-group dose to the regulatory

---

[5] The ICRP defines critical group in terms of dose. We use the ICRP terminology here to describe the concept as developed by the ICRP, and later adapt the concept to the risk framework.

limit is less than one-tenth, then the critical group should be considered homogeneous if the distribution of individual doses lies substantially within a total range of a factor of ten, or a factor of three on either side of the average. At ratios greater than one-tenth, homogeneity requires a smaller range.

3. The group be relatively small. The ICRP recommends that it should typically include a few to a few tens of persons. Normally a critical group would not consist of a single individual but rather a few tens of individuals. On the other hand, homogeneity implies that the group should not be too large.

In the context of an individual-risk standard, similar conditions would apply for the same reasons. Based on cautious, but reasonable, assumptions, the group would include the persons expected to be at highest risk, would be homogeneous in risk[6], and would be relatively small. The critical-group risk calculated for purposes of comparison with the risk limit established in the standard would be the mean of the risks of the members of the group.

More specifically, we recommend the following definition of the critical group for use with the individual-risk standard:

> The critical group for risk should be representative of those individuals in the population who, based on cautious, but reasonable, assumptions, have the highest risk resulting from repository releases. The group should be small enough to be relatively homogeneous with respect to diet and other aspects of behavior that affect risks. The critical group includes the individuals at maximum risk and is homogeneous with respect to risk.

---

[6] That is, the difference between the highest and lowest risk faced by individuals in the group should be relatively small. Should a radiation dose occur, however, it may affect only a few members of the group. This is the difference between risk (the probability of an adverse health effect) and outcome (a cancer that actually develops). Risk can be homogeneous, even when outcomes are quite diverse.

> A group can be considered homogeneous if the distribution of individual risk within the group lies within a total range of a factor of ten and the ratio of the mean of individual risks in the group to the standard is less than or equal to one-tenth. If the ratio of the mean group risk to the standard is greater than or equal to one, the range of risk within the group must be within a factor of 3 for the group to be considered homogeneous. For groups with ratios of mean group risk to the standard between one-tenth and one, homogeneity requires a range of risk interpolated between these limits.

This definition requires specifying the persons who are likely to be at highest risk. In the present and near future, these persons are real; that is, they are the persons now living in the near vicinity of the repository that lies in the direction of the flow of the ground water plume of radionuclides that would occur far in the future. The expected containment capability of an undisturbed repository at Yucca Mountain means, however, that no significant risks would likely arise until at least thousands of years in the future. At such times, it will be necessary to define hypothetical persons by making assumptions about lifestyle, location, eating habits, and other factors. ICRP recommends use of present knowledge and cautious, but reasonable, assumptions in making projections far into the future. These assumptions are part of the exposure scenarios[7] that must be defined as a basis for determining whether the repository performance is judged to comply with the standard. Exposure scenarios are discussed further in the next chapter.

For How Long?

As noted earlier, the current EPA standard contains a time limit of 10,000 years for the purpose of assessing compliance. There are three possible reasons for setting such a time limit. One would be to set a policy

---

[7] There are multiple release pathways from the repository, and each might have its own exposure scenario and critical group. However, only one of these critical groups will contain the person or persons that face the highest risk.

that beyond a set interval of time, it would not be necessary to protect public health. We will not address this reason, but only the other two, which have a technical basis.

The first technically based reason is the argument that beyond that limit the uncertainties in compliance assessment become too large. We consider this issue in Chapter 3, and conclude that assessment is feasible for many aspects of repository performance for much longer times and that the ultimate restriction on time scale is determined by the long-term stability of the fundamental geologic regime — a time scale that is on the order of 10^6 years at Yucca Mountain. In the case of human activity, as discussed in Chapters 3 and 4, there is no scientific basis for prediction of future states, and the limit of our ability to extrapolate with reasonable confidence is measured in decades or, at most, a few hundreds of years.

The other technically based reason for limiting the time of analysis is if there are likely to be no significant health effects after a specified time. In the case of Yucca Mountain, at least, some potentially important exposures might not occur until after several hundred thousand years. For example, the half-life of some of the radionuclides contained in the repository is millions of years, and for some scenarios the travel time of these materials to the accessible environment is in the range of tens of thousands to hundreds of thousands of years.

For these reasons, we believe that there is no scientific basis for limiting the time period of the individual-risk standard to 10,000 years or any other value. We recommend in Chapter 3 that compliance assessment be conducted for the time when the greatest risk occurs, within the limits imposed by long-term predictability of both the geologic environment and the distribution of local and global populations.

Indeed, the 10,000-year limitation might be inconsistent with protection of public health. For example, as noted in a previous National Research Council study," EPA's 10,000-year time limit, evidently adopted in USNRC's rationale, makes compliance rather easy. This we do not support because . . . we see no valid justification for this time limit . . . The USNRC-EPA calculational approach may seem to simplify licensing, but we do not understand how such an exercise can support the finding, required in licensing, that there be no unreasonable risk to the health and safety of the public" (NRC, 1983, at p. 236).

As described, we have recommended that the standard for individual risk should apply at times when the peak potential risks might occur. We recognize that there are significant uncertainties in the supporting calculations and that the uncertainties increase as the time at which peak risk occurs increases. However, we see no technical basis for limiting the period of concern to a period that is short compared to the time of peak risk or the anticipated travel time.

Nevertheless, we note that although the selection of a time period of applicability has scientific elements, it also has policy aspects that we have not addressed. For example, EPA might choose to establish consistent policies for managing risks from disposal of both long-lived hazardous nonradioactive materials and radioactive materials.

Another time-related regulatory concern can affect the formulation of the safety standard. This is based on ethical principles, and is the issue of intergenerational equity (Berkovitz, 1992; Holdren, 1992; Okrent, 1994). Whether and how best to be fair to future generations is an important societal question. Although current generations are assumed to have benefited from activities, such as electricity production or national defense programs that have caused radioactive wastes to accumulate, far future generations will not benefit directly, but might be exposed to risks when any radioactive materials eventually escape the proposed repository. In drafting standards, EPA should as a matter of policy address whether future generations should have less, greater, or equivalent protection.

The responsible institutions have considered the question of the protection to be afforded future generations. For example, in her presentation to us, Margaret Federline (USNRC, personal communication, May 27, 1993) spoke about a "societal pledge to future generations" that would "provide future societies with the same protection from radiation we would expect for ourselves." The IAEA document, *Safety Principles and Technical Criteria for HLW Disposal*, Safety Series 99, has as one objective the "responsibility to future generations." Under this responsibility to future generations, IAEA recommends that "the degree of isolation of high-level radioactive waste shall be such so there are no predictable future risks to human health or effects on the environment that would not be acceptable today." In this IAEA establishes that "[t]he level of protection to be afforded to future individuals should not be less than that provided today."

A health-based risk standard could be specified to apply uniformly over time and generations. Such an approach would be consistent with the principle of intergenerational equity that requires that the risks to future generations be no greater than the risks that would be accepted today. Whether to adopt this or some other expression of the principle of intergenerational equity is a matter for social judgment.

PROTECTING THE GENERAL PUBLIC

Earlier in this chapter, we recommend the form for a Yucca Mountain standard based on individual risk. Congress has asked whether standards intended to protect individuals would also protect the general public in the case of Yucca Mountain. We conclude that the form of the standards we have recommended would do so, provided that policy makers and the public are prepared to accept that very low radiation doses pose a negligibly small risk. This latter requirement exists for all forms of the standards, including that in 40 CFR 191. We recommend addressing this problem by adopting the principle of negligible incremental risk to individuals.

The question posed by Congress is important because limiting individual dose or risk does not automatically guarantee that adequate protection is provided to the general public for all possible repository sites or for the Yucca Mountain site in particular. As described in the previous section, the individual-risk standard should be constructed explicitly to protect a critical group that is composed of a few persons most at risk from releases from the repository. The standards are then set to limit the risk to the average member of that group. Larger populations outside the critical group might also be exposed to a lower, but still significant, risk. It is possible that a higher level of protection for this population represented by a lower level of risk than the one established by the standards might be considered.

For purposes of this discussion, the "general public" can be thought of as including global (hemispheric or continental) populations that might receive very small risks from repository releases, as well as local populations that lie outside the critical group but that might still be exposed to risks not much lower than those imposed on the critical group. The issues are different for these two types of populations, and we discuss them separately.

PROTECTING THE GLOBAL POPULATION

Radiation releases from a repository can in principle be distributed to a global, or other large and dispersed population, in several ways. For example, food contaminated by radionuclides could be shipped to regions far from the repository area, or contaminated ground water could enter a major river and the drinking water supplies that it serves. The global distribution of releases from a repository is assumed as the exposure scenario for the containment requirements in EPA's regulation 40 CFR 191. In the case of Yucca Mountain, there would be no releases to major rivers, and therefore the most likely pathways for global distribution are gaseous releases of carbon dioxide containing the radioactive isotope of carbon, ^{14}C, that eventually will escape from the waste canisters, or by widespread distribution of foodstuffs grown with contaminated water.

In general, the risks of radiation produced by such wide dispersion are likely to be several orders of magnitude below those to a local critical group. As noted earlier in this chapter, however, the "linear hypothesis" implies that even very small increments to background doses might cause effects from cancer induction in the same ratio ($5x10^{-2}$/Sv) as larger doses. Using the linear hypothesis to calculate the effects of very low doses on large populations requires multiplying this factor by the cumulative dose imposed on populations numbered in the trillions over the life of the repository.

There are, however, important cautions to be noted with this procedure. With respect to small increments to natural background radiation levels, the BEIR V report (NRC 1990a) states that:

> Finally, it must be recognized that derivation of risk estimates for low doses and dose rates through the use of any type of risk model involves assumptions that remain to be validated. At low doses, a model dependent interpolation is involved between the spontaneous incidence and the incidence at the lowest doses for which data are available. Since the committee's preferred risk models are a linear function of dose, little uncertainty should be introduced on this account, but departure from linearity cannot be excluded at low doses below the range of observation. Such departures could be in the direction

> of either an increased or decreased risk. Moreover, epidemiologic data cannot rigorously exclude the existence of a threshold in the millisievert dose range. Thus the possibility that there may be no risks from exposures comparable to external natural background radiation cannot be ruled out. At such low doses and dose rates, it must be acknowledged that the lower limit of the range of uncertainty in the risk estimates extends to zero.[8]

The doses to global populations involved in gaseous release from Yucca Mountain are likely to be well below the mSv range noted in BEIR V. For example, let us assume that the repository inventory of 91,000 Ci (3.37×10^{15} Bq) (Wilson et al., 1994) of ^{14}C is released into the air over 10,000 years. Using EPA's dose conversion factor 1.1×10^{-10} Sv/Bq (EPA, 1992), the population dose over 10,000 years would be 3.7×10^5 person-Sv, or an average of 37 person-Sv/year over the 10,000-year period (Nygaard et al., 1993). Assuming that the ^{14}C is well mixed with air over the globe, and for an average global population of 12 billion people during this period, the corresponding average individual dose rate is 3.1×10^{-9} Sv/yr (3.1×10^{-4} mrem/yr). For comparison, the dose set by EPA in 40 CFR 191 is 1.5×10^{-4} Sv/yr (15 mrem/yr), and this is the limit to be applied for the persons likely to receive the highest doses from the repository. Therefore, there is great uncertainty about the number of health effects that would be imposed on the global population because of the difficulties in interpreting the risks associated with such small incremental risks from ^{14}C releases at Yucca Mountain.

NEGLIGIBLE INCREMENTAL RISK

To address scenarios of widespread but extremely low-level doses, the radiation protection community has introduced the concept of negligible individual dose. The negligible individual dose is defined as a level of effective dose that can, for radiation protection purposes, be dismissed from consideration. NCRP has recommended a value of 0.01

---

[8] In this paragraph "low doses" applies to very small increments to the dose from the natural background.

mSv/yr (1 mrem/yr) per radiation source or practice (NCRP, 1993), which currently would correspond to a projected risk of about 5 x 10^{-7}/yr for fatal cancers, assuming the linear hypothesis. In its considerations, NCRP decided on this level of dose or risk taking into account risk in relation to:

1. Natural risk of the same health effects;
2. Risk to which people are accustomed;
3. Estimated risk for the mean and variance of natural background radiation exposure levels;
4. Perception of, and behavioral response to, risk levels; and
5. Difficulty in detection and measurement of dose and health effects.

Others over the years have advocated the use of a negligible dose or risk level (Comar, 1979; Eisenbud, 1981; Schiager et al., 1986)[9]. The general consensus of these authorities was that a negligible value would be useful in many applications. Federal and state approaches for the regulation of chemical carcinogens are in keeping with this view, which generally take a 10^{-6} lifetime risk as an acceptable level (Travis et al., 1987; EPA, 1991), as are the exposure limits for radioactive waste adopted by most nations in the Organization for Economic Cooperation and Development (OECD) (Dejonghe, 1993). The Federal German Radiation Protection Commission, for example, has recommended ignoring individual doses of less than 0.003 mSv per year (Smith and Hodgkinson, 1988).[10]

We believe that the concept of a negligible incremental dose can be extended to risk and can be applied to Yucca Mountain. Defining the level of incremental risk that is negligible is a policy judgment. We suggest the risk equivalent of the negligible incremental dose recommended by the NCRP as a reasonable starting point for developing consensus in a rulemaking process. For example, the average dose to a member of the global population from exposure to ^{14}C from the repository

---

[9] Where authors use "negligible dose" or "negligible risk" the terms should be understood as increments to the unavoidable background radiation. In life, there is no zero dose and no zero risk.

[10] Note that this is equivalent to an annual risk of fatal cancer of about 1.5x10^{-7}/yr.

is estimated to be about 3 x 10^{-9} Sv/yr, corresponding to a risk of fatal cancer of 1.5 x 10^{-10}/yr or about 10^{-8} per lifetime. As indicated earlier, NCRP has recommended a negligible incremental dose that corresponds to a risk of 5 x 10^{-7}/yr (NCRP, 1993). Therefore, if the NCRP recommendation were adopted, the effects of gaseous ^{14}C releases on individuals in the global population would be considered negligible.

PROTECTING LOCAL POPULATIONS

Persons in some populations outside the critical group might be exposed to risk from repository releases in excess of the level of negligible incremental risk. As individuals, these persons would be (by definition and in practice) exposed to less risk than the risk limit established by the standard for the critical group. If many persons were exposed to this individual risk, however, the total number of health effects that could occur might be relatively large, particularly if integrated over a very long period of time.

We know of no analysis that has addressed the spatial distribution of radiation doses and risks near Yucca Mountain at the distant future times when individual doses and risks would be at their maximum. It should be feasible to determine a spatial distribution of potential concentrations in ground water or air and a spatial distribution of individual doses and risks, employing the same types of exposure assumptions used for calculating doses and risks to members of a critical group (see Chapter 3). However, the total number of fatal cancers cannot be known without knowledge of the number of future persons residing in the Yucca Mountain vicinity. This number is obviously unknowable. Even if EPA were to define it arbitrarily through a rulemaking process, comparing the total population risk against some defined figure-of-merit in order thereby to decide on whether to accept or reject a repository seems too arbitrary to be useful.

Population-Risk Standard

As an example of the difficulty of framing an absolute population-risk standard, we considered normalizing the population risks as a means to avoid the difficulty of not having a technical basis for knowing the total population at risk. Such a regulatory scheme might require that the integrated population risk over a given period (one generation, for example) be limited to some fractional risk in the affected population. A specific hypothetical example would be to require that the integrated population risk must produce fewer than *x* health effects per N people during a defined interval of time.

Framed this way, however, the standard looks very much like an individual-protection standard: each person outside the critical group would have an individual lifetime risk limited to *x*/N. As a matter of policy, it is certainly legitimate to desire to protect a smaller group (the critical group) by limiting individual risk to a certain level, and also to protect a larger group (the nearby population) with a different but still meaningful risk limit. However, this approach is not a collective-risk protection scheme — it is merely a two-tiered individual-risk protection scheme.

Spatial Gradient in Risk

An alternative approach that does have a technical basis is consideration of the spatial distribution of individual risks near the critical group, at the distant future time when the critical-group risk is highest. Such a spatial distribution has a technical significance because it depends on the characteristics not only of the Yucca Mountain physical site but also of the waste form and the engineered and geologic barriers of the repository design.

Furthermore, a risk distribution with a steep spatial gradient — that is, a distribution in which the individual risks become smaller relatively quickly with increasing distance from the location of the highest individual risks — seems obviously preferable to a distribution with a more gradual spatial gradient, all other things being equal. This is because a steeper spatial gradient implies smaller integrated population risks than does a more gradual gradient for the same spatial distribution of population.

This observation cannot provide information for discriminating between an "acceptable" repository and an "unacceptable" one without an acceptable level of risk for comparison purposes. However, we have not been able to identify a technically based figure-of-merit that could be used to judge the compliance acceptability of a given spatial risk gradient. To use the gradient in an absolute sense, one is faced with not only selecting a time interval of concern, which is arbitrary, but also defining the future nearby population. For the simpler task of adequately characterizing the exposure scenarios leading to calculation of risks to a critical group, we have concluded that a feasible procedure can be developed using known distributions of physical and chemical parameters and defensible assumptions on lifestyles; in other words, there is a reasonable technical basis for a critical-group calculation. For identifying the size, the distribution and the varied lifestyles of a larger population, more assumptions of greater uncertainty would be required. The resulting data for a risk assessment would become so arbitrary that no adequate decision basis would result. We therefore conclude that there is no technical basis for establishing a population-risk standard that would limit the risk to the nearby population for a Yucca Mountain repository.

PREFERRED FORM OF THE STANDARD

Although we have couched the discussion of the last two sections in terms of an individual-risk standard, we noted in an earlier section of this report that there are several possible forms of standard that could be used. We end this chapter by explaining why we conclude that the individual-risk form has scientific advantages over the others.

Release Limits. It is possible to state the standard in terms of a limitation on the amount of radionuclides crossing an imaginary boundary that encloses the repository. The limit generally would be placed on cumulative release over a specified time period. This is the approach used by EPA in 40 CFR 191, which relies primarily on a table of maximum allowable cumulative radioactive releases to the accessible environment for a period of 10,000 years.

A release limit has the appearance of simplicity because it focuses on the amount of radionuclides released from the repository across some specified boundary. This form of standard does not provide any information about how these releases affect public health, however, and so is incomplete unless coupled with a calculation of individual (or population) risk (or dose or health effects). If one is interested in this information on public health for a specific site, it is good scientific practice to incorporate specific data about the site into the calculation. If that is done, essentially all of the calculations described in Chapter 3 are required. The advantage of our recommendation is that these calculations are to be done using a methodology approved by a rulemaking, with all calculations explicit to the public. Hence, we conclude that a release limit for a site-specific standard does not reduce scientific complexity or uncertainty. Without calculations of dose or risk, a release standard appears arbitrary.

Other than the appearance of simplicity, there seem to be no other advantages to a release-limit form of the standards. It does not produce information that is easy to understand or to compare with other risks. Note that no other standard listed in either Table 2-3 or 2-4 is expressed as a release limit.

A population standard[11], such as the one that appears to be the basis for the release limit in 40 CFR 191, establishes a total number of health effects permitted over some time period — 1,000 in 10,000 years, in the case of 40 CFR 191. This form of standard does not provide a basis for assessing the risk to the individuals in the critical group, or for local populations nearby. Therefore, a population standard alone is insufficient to protect the population most at risk and, probably for this reason, 40 CFR 191 contains a parallel individual standard.

Also, as discussed earlier in this chapter, assessing compliance with a standard designed to protect the global population involves highly uncertain calculations because of the extremely low incremental doses to which large numbers of persons may be exposed. We have recommended the use of the concept of negligible incremental risk to individuals as a preferable way of dealing with these uncertainties at the outset.

An *individual standard* is needed, however, and the issue is whether to state it in terms of dose, health effects, or risk. In Section 801, Congress directs EPA to use individual dose. As mentioned above, we recommend using the risk form for the following reasons:

1. A risk-based standard would not have to be revised in subsequent rulemaking if advances in scientific knowledge reveal that the dose-response relationship is different from that envisaged today. Such changes have occurred frequently in the past, and can be expected to occur in the future. For example, ongoing revisions in estimates of the

---

[11] Or, equivalently, a cumulative dose standard.

radiation doses received by atomic bomb survivors of Hiroshima and Nagasaki may significantly modify the apparent dose-response relationships for carcinogenic effects in this population, as have previous revisions in dosimetry (see Straume et al., 1992).

2. Risks to human health from different sources, such as nuclear power plants, waste repositories, or toxic chemicals, can be compared in reasonably understandable terms. Doses or releases have to be stated in radiation units Sieverts or Becquerels that are not easily understood by the general public and that can only be compared conveniently with other sources of radiation or radioactivity.

Although we recommend a risk-based standard rather than the dose-based standard in Section 801, they are closely related. We define risk as the expected value of the probabilistic distribution of health effects. The distribution of health effects is derived from a distribution of dose and the expected health effects per unit dose.

Consequently, in answer to congressional question No. 1, we believe that a health-based individual standard will provide a reasonable standard for protection of the general public. However, we recommend that this be a risk-based, rather than a dose-based standard.

3

ASSESSING COMPLIANCE

INTRODUCTION

In the preceding chapter, we described our conclusion that the form of a Yucca Mountain standard should be based on limiting individual risk as measured by the average risk to individuals in a critical group. This group is defined as being composed of persons likely to be at highest risk from radionuclides released from the repository. Our judgment is that limiting individual risk in this way is also likely to provide adequate radiological protection for all relevant populations that might be exposed to radiation from radionuclides released from the proposed repository at Yucca Mountain (see Chapter 2). The period over which this level of protection should be assessed should extend over the period of duration of hazard potential of the repository, that is, until the time at which the highest critical group risk is calculated to occur, within the limits imposed by the long-term stability of the geologic environment at Yucca Mountain, which is on the order of 10^6 years.

In this chapter, we discuss the analyses that must be undertaken to judge compliance with such a standard. Important questions to be answered are:

1. Whether the scientific understanding of the relevant events and processes potentially leading to releases is sufficient to allow a quantitative estimate of future repository behaviors.
2. Whether adequate analytical methods and numerical tools exist to incorporate this understanding into quantitative assessments of compliance.
3. Whether the current scientific understanding and analytic methods are sufficient to evaluate performance with sufficient confidence to assess compliance over the long time periods required.
4. Whether the results of the analyses required to assess repository performance can be combined into an estimated

risk for comparison with the standards in the licensing process. In particular, the estimated risk is defined as the mean risk of members in the critical group. Risk is defined as the expected value of the probabilistic distribution of health effects experienced by an individual member of the critical group.

The main tool used to assess compliance is quantitative performance assessment, which relies upon mathematical modeling. We have evaluated the degree of confidence that can be placed today in such assessments. We have also made a systematic analysis of the application of this methodology to the Yucca Mountain site. Based on these analyses, we conclude that:

1. For those aspects of repository and waste behavior that depend on physical and geologic properties and processes, enough of the important aspects can be known within reasonable limits of uncertainty, and these properties and processes are sufficiently understood and stable over the long time scales of interest to make calculations possible and meaningful. These properties and processes include the radionuclide content of the waste (which changes over time due to radioactive decay), the influx of water through the site and its effect on waste package integrity and other engineered barriers, the migration of wastes to ground water after waste packages have lost their integrity, and the subsequent dispersion and migration of wastes in ground water. While these factors cannot be calculated precisely, we believe that there is a substantial scientific basis for making such calculations, taking uncertainties and natural variabilities into account, to estimate, for example, the concentration of wastes in ground water at different locations and the times of gaseous releases.

   One critical gap in our understanding is with respect to future human behavior. Since there is no scientific basis for predicting human behavior, we recommend that policy decisions be made to specify default (or reference) scenarios to be used to incorporate assumed future human behavior into compliance assessment calculations.

2. Available mathematical and numerical tools are neither perfect nor complete. Nevertheless, the currently available tools plus additional tools that we believe can be developed as part of the standard-setting and compliance assessment efforts, or through other research, should be adequate for the analyses required to evaluate repository performance.
3. So long as the geologic regime remains relatively stable, it should be possible to assess the maximum risks with reasonable assurance. The time scales of long term geologic processes at Yucca Mountain are on the order of 10^6 years. Other processes that operate on short time scales, such as seismic activity, can also be accommodated in performance assessment if the maximum risks associated with these processes depend more on whether an event is likely to occur (at any time) than on the specific timing of the event.
4. Established procedures of risk analysis should enable the combination of the results of all repository system simulations into a single estimated risk to be compared with the standard. (Human intrusion is excluded from such a combination. See Chapter 4.) An element of judgment is contained in many of the conceptual assumptions to be made, and those assumptions, methods, and the reference data will have to be specified. Similarly, reference exposure scenarios must be established clearly. This transparency in the use of assumptions is critical to evaluating the calculated risk.

Because some readers might be unfamiliar with the technical aspects of a repository performance assessment, it is appropriate to provide an overview of the methodology, as we do in Part I of this chapter. We then consider the scientific basis for making an assessment of Yucca Mountain. We have found it useful to separate this evaluation into two parts, one dealing with the physical properties and geologic processes relevant to the behavior of the wastes and the other with those aspects of performance assessment that deal with assumptions about where and how people live, how they might be exposed through the food and water they consume, and other factors that could affect exposures to radioactive wastes. We shall refer to this latter collection of factors that must be considered as exposure scenarios. The reason for separating these two elements of performance assessment is that the nature of calculations in each is substantially different. We discuss these in Parts II and III.

PART I: OVERVIEW OF PERFORMANCE ASSESSMENT

Any standard to protect individuals and the public after the proposed repository is closed would require assessments of performance at times so far in the future that a direct evaluation of compliance (for example by physical monitoring of system behavior) is out of the question. The only way to evaluate the risks of adverse health effects and to compare them with the standard is to assess the estimated potential future behavior of the entire repository system and its potential impact on humans. This procedure, involving modeling of processes and events that might lead to releases and exposures, is called performance assessment. It involves computer calculations using quantitative models of physical, chemical, geologic, and biological processes, taking uncertainties into account.

Modeling repository performance is a challenging task because the rates of geochemical transformation and transport of the radionuclides are generally very slow and the times at which points distant from the repository become significantly affected by radionuclide releases will be in the far future. Thus, to assess these effects requires projection of geochemical, hydrodynamic, and other processes over long time periods within rock masses whose properties are imperfectly known. Factors describing how humans can be exposed to radionuclides from the wastes are even more imperfectly known and these factors, including the future state of technology and medicine, might be more changeable over time than are the physical processes.

Reasonable Confidence

One possible response to these difficulties is to conclude that they render any assessments of the ultimate fate of the waste materials too uncertain to be useful. However, we believe that such analyses do provide information for judging the quality of a disposal site. Even if the uncertainties involved are large, some options for the disposition of the wastes can clearly be shown to result in worse consequences than other options would produce.

The results of compliance analysis should not, however, be interpreted as accurate predictions of the expected behavior of a geologic repository. No analysis of compliance will ever constitute an absolute proof; the objective instead is a reasonable level of confidence in analyses that indicates whether limits established by the standard will be exceeded. Both the USNRC and EPA have explicitly recognized this objective. For example, EPA states in 40 CFR 191 that "unequivocal numeric proof of compliance is neither necessary nor likely to be obtained." In regulation 10 CFR 60, USNRC acknowledges that "it is not expected that complete assurance that [performance objectives] will be met can be presented." The USNRC requires instead "reasonable assurance, making allowances for the time period, hazards, and uncertainties involved." EPA's required level of proof in 40 CFR 191 is "reasonable expectation."

Time scale

One commonly expressed concern regarding the performance assessment modeling is that it requires simulating performance at such distant times in the future that no confidence can be placed in the results. Of course, the level of confidence for some predictions might decrease with time. This argument has been used to support the concept of a 10,000 year cutoff (DOE, 1992). We do not, however, believe that there is a scientific basis for limiting the analysis in this way.

One of the major reasons for selecting geologic disposal was to place the wastes in as stable an environment as many scientists consider possible. The deep subsurface fulfills this condition very well (NRC, 1957). In comparison with many other fields of science, earth scientists are accustomed to dealing with physical phenomena over long time scales. In this perspective even the longest times considered for repository performance models are not excessive. Furthermore, even changes in climate at the surface would probably have little effect on repository performance deep below the ground. We recommend calculation of the maximum risks of radiation releases whenever they occur as long as the geologic characteristics of the repository environment do not change significantly. The time scale for long-term geologic processes at Yucca Mountain is on the order of approximately one million years. After the geologic environment has changed, of course, the scientific basis for performance assessment is substantially eroded and little useful information can be developed.

Because there is a continuing increase in uncertainty about most of the parameters describing the repository system farther in the distant future, it might be expected that compliance of the repository in the near term could be assessed with more confidence. This is not necessarily true. Many of the uncertainties in parameters describing the geologic system are due not to temporal extrapolation but rather to difficulties in spatial interpolation of site characteristics. These spatial difficulties will be present at all times. Accordingly, even in the initial phase of the repository lifetime, a compliance decision must be based on a reasonable level of confidence in the predicted behavior rather than any absolute proof. Under some circumstances, use of a shorter period for analysis could in fact introduce additional uncertainties into the calculation. For example, uncertainties in waste canister lifetimes might have a more significant effect on assessing performance in the initial 10,000 years than in performance in the range of 100,000 years.

Probabilistic Analysis of Risk

To judge compliance against a risk-based standard of the type proposed, a risk analysis including treatment of all scenarios that might lead to releases from the repository and to radiation exposures is, in principle, required. To include them in a standard risk analysis, all these scenarios need to be quantified with respect to the probabilities of *scenario occurrence* and the probability distribution of their *consequences* to humans, such as health effects of radiation doses. In subsequent sections we specifically note that for some events or processes either the probability of occurrence or the estimated consequences become very difficult to specify with confidence. Events caused by human activity are usually of this type. Incorporation of such events or processes into the formalized risk analysis sometimes is not justified on a scientific basis. Instead, how to deal with these events should be decided as a matter of policy.

This approach implies a departure in part from common analytical techniques to assess risks and the introduction of more pragmatic procedures needed to provide an adequate decision basis. It is important, therefore, that the "rules" for the compliance assessment be established in advance of the licensing process; that is, that the scenarios that might be excluded from the integrated risk analysis be identified. Human intrusion is an example of one scenario that we judge to be not amenable to incorporation in the risk assessment framework; this is discussed further in Chapter 4.

We believe that performance assessment using numerical models of physical and chemical processes and quantitative estimates of probabilities is the key approach to assessing compliance. However, the confidence that can be placed in such analyses is also a key part of the compliance issues. To some extent, this degree of confidence can be quantified, for example, by performing rigorous uncertainty analyses that propagate uncertainties in parameter values through the analysis to produce estimates of uncertainties in estimated risks. Uncertainties due to modeling approaches can also be assessed by comparing the results of assessments using various alternative models, or by comparing model results with data collected in experiments or in observations. In other cases, less rigorous but useful evidence of the adequacy of models or data can be obtained by, for example, comparisons with relevant natural analog systems.

A final, important point to note is that performance assessments of the type summarized above are not likely to be performed only on a single occasion preparatory to licensing. Assessments will likely be performed iteratively during system design, construction and operation of a geologic repository, and finally at the time the repository is sealed, following decades of experience in which additional data on the performance of system components can be gathered.

QUANTITATIVE CALCULATION OF REPOSITORY PERFORMANCE

In this section, we summarize general aspects of performance assessment modeling and sources of uncertainty in the modeling process before moving in subsequent sections to issues more specific to Yucca Mountain. The main thrust of performance assessment involves developing a quantitative understanding of system behavior, assembling a sufficient database of parameters describing the system, and producing simulations of possible future system behavior allowing as fully as possible for uncertainties in understanding or in databases. Figure 3.1 schematically illustrates the generic modeling process described in more detail below.

Figure 3.1 The Basic Steps in Performance Assessment

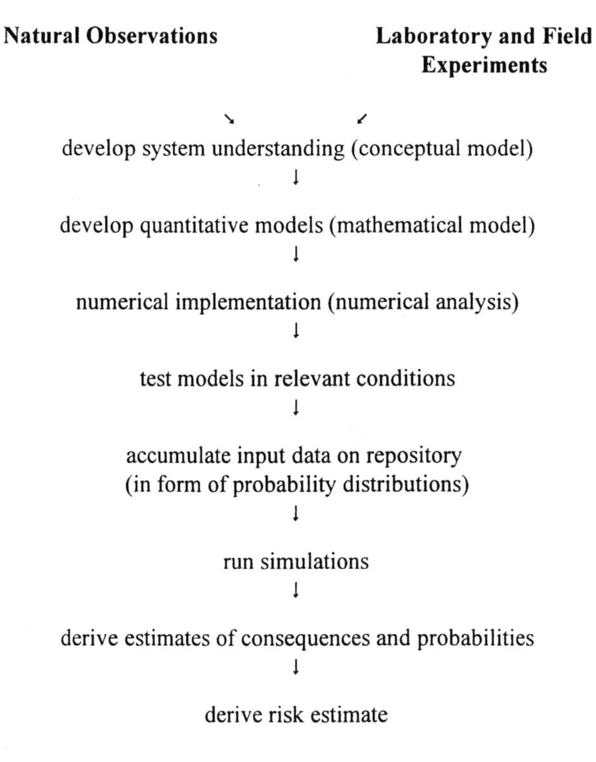

Elements of Performance Assessment

Conceptual model

The conceptual model reflects the scientists' understanding of how the important aspects of the system work. It answers questions such as:

What are the limits of the system? What are the geometry and composition of the system? What are the significant physical processes? It is the conceptual model that dictates the selection of the mathematical formalisms that enable quantitative calculations to be performed.

One special type of conceptual model frequently employed in performance assessment is the scenario. In this context, a scenario means a description of how radionuclides might migrate from the repository and affect humans. For example, "the wastes are dissolved in ground water, which is transported by natural processes to an agricultural area, where it is pumped out of the ground and used to irrigate crops and ingested by humans" is a possible scenario for the Yucca Mountain repository. Quantitative performance assessment based on this scenario would then have to employ detailed conceptual models of release and transport processes specifying, among other things, how and where the ground water flows and exposure scenario models specifying where farmers live, what technologies they use and their patterns of consumption of food and water. The scenario thus constitutes a kind of master conceptual model that guides the selection of more detailed and specific conceptual models for each step of the process.

The conceptual models are potentially the source of the most significant uncertainties regarding the outcome of the analysis. If the nature of the system has not been properly assessed, or the most important processes have not been included in the conceptual model, the mathematical model based on the conceptual model will not properly simulate the behavior of the system regardless of how adequately the other elements of the analysis might be quantified.

Inadequacies in conceptual models are a particularly worrisome aspect of the performance assessment process because a major error could invalidate the entire exercise, yet be difficult or impossible to detect. Although, it is important to realize that this limitation is an aspect of all human problem-solving activities, it is particularly important for radioactive waste repository performance assessment computations because of their long-term considerations. The best way to guard against errors of this nature is to provide for multiple, rigorous, independent reviews of conceptual models that are clearly documented and widely disseminated.

Mathematical model

By mathematical model we mean the mathematical relationships that are used to describe the physical system quantitatively. The system of equations that is incorporated in the mathematical model usually represents a simplification of the selected conceptual model. Mathematical simplification might be required because it is not possible to find adequate descriptions of all the phenomena considered important, or because incorporation of all relevant equations would result in a mathematical system too cumbersome to solve, or because the data available do not justify the most complete description of the system that might be possible. Mathematical simplifications reduce the realism of the outcome of the model, but the degree to which the results are affected can be assessed by means of mathematical techniques, such as sensitivity analyses of numerical results.

Numerical analysis

Most mathematical models consist of sets of coupled differential equations. For the cases of interest to performance assessment, it is often difficult to solve such complex systems of equations analytically, or exactly, in which case approximate numerical methods are employed. Selection of appropriate numerical methods is important because more efficient numerical techniques can permit more complex (and thus, presumably, more realistic) physical models to be solved, and because inappropriate numerical schemes can introduce significant errors into results. However, numerical inaccuracies are rarely a major source of error in properly conducted modeling because well-established methods exist for assessing the accuracy of numerical schemes. Further, if one approach is found to introduce unacceptable error, it can either be replaced or modified to achieve the desired accuracy.

Model parameters

Physical and chemical models require the specification of the physical properties of the system to be modeled. These properties are referred to as parameters. The parameters are represented by numerical functions or values in the mathematical models. Models of the type commonly used in performance assessment describe the behavior of the system as a function of both space and time. Spatially heterogeneous models of systems incorporate the spatial variations of the parameters throughout the physical domain that is being modeled. The need to provide numerical values for parameters is another source of uncertainty in mathematical modeling. It is a goal of geologic disposal of nuclear wastes to emplace them in an environment that is deep, remote, and difficult to access. These same repository properties make it difficult to obtain data on the spatial variations of physical parameters in the system. Furthermore, the very procedures necessary to collect the data, such as drilling exploratory holes to extract samples of rock might compromise the integrity of the geologic barriers.

Boundary conditions

Performance assessment models have both spatial and temporal boundaries, that is, times of the beginning and ending of simulations. In general, both mass and energy can flow across these boundaries. Thus, to perform model calculations it is necessary to specify the conditions at the spatial and temporal boundaries (the model calculates parameter values within the model domain). Specification of the "boundary conditions" is subject to many of the same types of uncertainty that are involved in specifying parameter values, and they are usually dealt with in a similar fashion.

In general, spatial boundary conditions of regional scale subsurface flow models are considered to be constant over time. There is at least one important exception to this generalization. The upper boundary to the geologic environment around the repository is the atmosphere. The average of atmospheric conditions is the climate, and it is well known that climate can vary significantly over geologic periods of time. Although the typical nature of past climate changes is well known, it is obviously impossible to predict in detail either the nature or the timing of future climate change. This fact adds to the uncertainty of the model predictions.

During the past 150,000 years, the climate has fluctuated between glacial and interglacial status. Although the range of climatic conditions has been wide, paleoclimatic research shows that the bounding conditions, the envelope encompassing the total climatic range have been fairly stable (Jannik et al., 1991; Winograd et al., 1992; Dansgaard et al., 1993). Recent research has indicated that the past 10,000 years are probably the only sustained period of stable climate in the past 80,000 years (Dansgaard et al., 1993). Based on this record, it seems plausible that the climate will fluctuate between glacial and interglacial states during the period suggested for the performance assessment calculations. Thus, the specified upper boundary, or the physical top boundary of the modeled system, should be able to reflect these variations (especially in terms of ground water recharge).

Treatment of Uncertainty

The description above has emphasized sources of uncertainty in performance assessment. Uncertainties in scenario and detailed conceptual models are among the most important, but are difficult to quantify. Parameter uncertainty is also obviously important, but can be more rigorously treated. Compliance with a health standard can be judged acceptable only if the calculated behavior — even allowing for uncertainties in the analyses — is acceptable. Hence, the standard must require that estimates of technical uncertainties be provided even if it does not explicitly state in advance the permissible level of uncertainty. Some of the main issues in treating uncertainty are discussed below.

Probabilistic modeling

A number of statistical approaches exist to account for the effects of uncertainty in modeling the transport of radionuclides. A method used to help implement statistical distributions of a parameter in performance assessment is the Monte Carlo method. In this method, data on the frequency distributions of parameter values are sampled to provide input to the equations. These distributions are used to describe parameters where there is inherent variability or where the precise value is uncertain. The model is then run a number of times using parameter values randomly selected from the specified distributions. When a sufficient number of simulations have been performed, the statistics of the results are used to estimate the uncertainty imparted to the result by the uncertainty in the input parameters (Henley and Kumamoto, 1992).

The main problem for technical analyses of this type for compliance purposes might be developing consensus on the input statistical distributions of parameter values for the assessments. Because of the requirements for spatial resolution or the infrequency of particular events, deriving the distributions from measurement programs or from observations might not be feasible for defining the parameter distributions. This means that a large element of informed judgment will often be involved. A further drawback of complex probabilistic modeling is that the results are not very transparent or easily understood.

Bounding estimates

Analyses using pessimistic scenarios and parameter values are more easily understood than Monte Carlo analysis. The results of these conservative calculations are then no longer estimates of likely behavior but rather bounding estimates. Bounding estimates can be criticized for compounding conservative assumptions, since they can easily produce consequences that are highly improbable. On the other hand, if compliance can be shown with a bounding estimate, then there is no need for a more complex analysis. Bounding estimates can thus be very useful, but care should be given as to how one could combine the robust, bounding-estimate type of assessment with a probabilistic analysis.

Alternative conceptual models

In the case of uncertainty arising from the choices in conceptual models, even more difficult questions arise. It is sometimes tempting to treat alternative, physically exclusive conceptualizations of a particular process, such as unsaturated flow together in a combined probabilistic analysis by allocating to each concept some, possibly arbitrary, probability of being correct. This approach is hard to defend, although it is being used by some groups that are analyzing repositories. Alternatives include separate treatment as two scenarios, agreement on the most likely case, or concentrating on the more conservative case. In any event, explicit recognition of differences in expert opinions is unavoidable.

When all reasonable steps have been taken to reduce technical uncertainties by, for example, performing site characterization and material testing programs, there still remains a residual, unquantifiable uncertainty. It can never be totally ruled out that the best analytic conclusions might be affected by some hitherto unknown or overlooked process or event. This is not a situation unique to waste disposal; it occurs in other licensing arenas. The only defense against it is to rely on informed judgment. The formulation of any disposal standard and of corresponding compliance requirements should explicitly acknowledge that this is the case. Unfulfillable expectations can thus be avoided and a more defensible approach to licensing procedures might be possible.

Summary

This section has described a methodology for assessing the performance of nuclear waste repositories. The description has emphasized sources of uncertainty in the analysis, the most important of which are uncertainties in scenarios, detailed conceptual models, and parameters. These assessments can provide both analyses of the future performance of the repository and estimates of the uncertainty in the performance assessments. Further, both of these results have additional uncertainty due to factors, such as conceptual model uncertainty that might not have been properly quantified or for which quantification is not possible. The issue is whether the methods and data available today are capable of producing assessments of behavior (or else bounding estimates) adequate for indicating whether standards can be met.

This question has been addressed in international circles. Following a major conference on safety assessment in 1990, a Collective Opinion was prepared by the Radioactive Waste Management Committee of the Nuclear Energy Agency of the Organization for Economic Cooperation and Development. The general conclusions drawn were, first, that appropriate performance assessment tools are currently available for producing results of the quality required for a decision on compliance and, second, that the final quality of the results is restricted primarily by the availability of site-specific data for the analyses. We concur with these two conclusions.

PATHWAYS AND PROCESSES FOR PERFORMANCE ASSESSMENT AT YUCCA MOUNTAIN

We now turn to more specific consideration of factors that would enter into compliance calculations for a repository at Yucca Mountain. Comparisons of potential repository performance to a standard expressed in the form of individual risk require estimating the probabilistic distribution of doses to a critical group as well as a conversion from doses to health effects. Estimating the probabilistic distribution of doses requires identification of the potential pathways of radionuclides from the repository to the biosphere, which comprises the air, water, food and other components of the landscape that are accessible to humans as well as the humans themselves; estimates of the concentrations that will be present in air, water, food, and other materials with which humans might come into contact; and estimates of the probabilities that humans will be exposed to contaminated air, water, food, or other materials leading to a radiation dose.

The major pathways from a repository at Yucca Mountain to humans are illustrated schematically in Figure 3.2. In this figure, major reservoirs that can contain radionuclides at various times after closure of a repository are shown as rectangles. These include (1) the canisters or other waste forms in the repository horizon; (2) the backfill, disturbed rock and other materials of the near-field zone in the vicinity of the waste; (3) the rock, air and water of the unsaturated zone (rock and pores above the water table); (4) the local atmosphere above Yucca Mountain; (5) the world atmosphere; (6) the water table aquifer immediately beneath the repository; (7) the aquifer downgradient of the repository (away from the repository along the direction of ground water flow) from which water may be withdrawn via wells for human use; and (8) the regional discharge zone of the ground water flow system where water exits the ground as discharge to surface water bodies or through evapotranspiration. It should be noted that in most cases these reservoirs are not physically distinct at their boundaries but rather form a continuum with the next reservoir in the pathway.

Figure 3.2 Schematic illustration of the major pathways from a repository at Yucca Mountain to humans

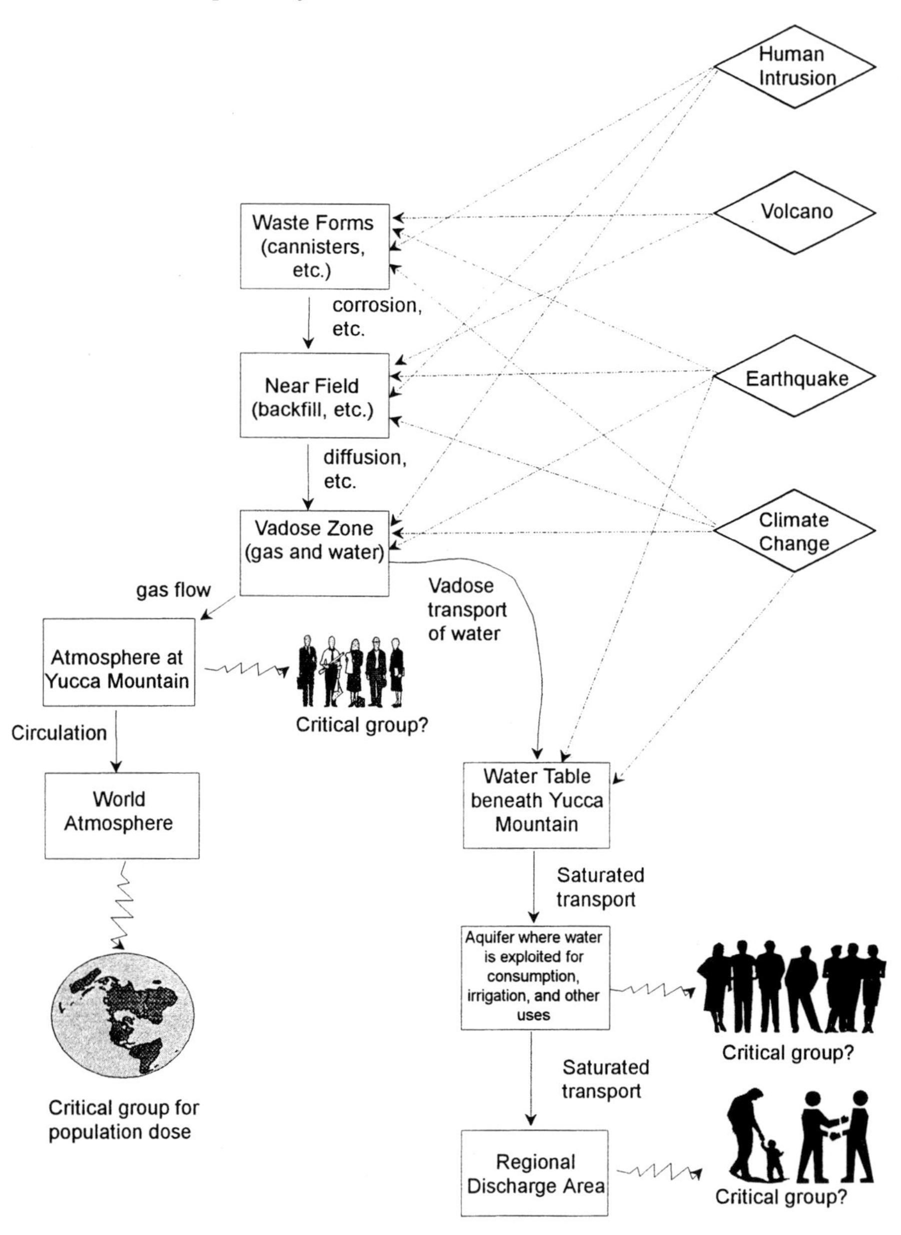

Solid arrows drawn on the figure from one box to another represent the processes by which radionuclides are transported from one reservoir to the next. The relative lengths of these arrows are meant to suggest, in a very qualitative sense, the relative times involved in these transport processes. For example, release of gaseous radionuclides from waste packages and their transport through the unsaturated zone to the atmosphere above Yucca Mountain is thought to be a relatively rapid process compared to dissolution of radionuclides and their transport in solution through the unsaturated zone to the water table.

Exposure pathways from the atmosphere or from ground water to humans are represented by jagged arrows. These arrows represent not only factors that affect human exposures, such as geographic location and eating and drinking habits, but also the human response to radiation doses. Releases to the atmosphere directly above or adjacent to Yucca Mountain can cause exposure by inhalation to the people who might be present in the immediate vicinity and who constitute a potential critical group. Atmospheric circulation, which will dilute concentrations many orders of magnitude from those at the mountain, can lead to worldwide exposures of the world population. Exploitation of ground water by some potential critical group downgradient from the repository can lead to exposure via food, water, or other contact with contaminated water. If radionuclides are transported through the entire ground water flow system to the regional discharge area, they or their radioactive or stable daughters might accumulate in soil, water, plants, leading to possible exposures in yet another potential critical group.

Several gradual and episodic natural processes, specifically global-climate change, volcanic eruptions, and seismic activity, have the potential to modify the properties of the reservoirs and the processes by which radionuclides are transported through these reservoirs to the biosphere. These are shown on the figure as diamonds connected by dashed lines to the reservoirs upon which they are likely to have the most significant effects. Intrusion by humans into the repository also has the potential to modify properties of the repository as well as of the near-field and unsaturated zones in the vicinity of the intrusion. Human intrusion is represented as a diamond similar to those for the natural processes that could modify repository performance. Human intrusion could also lead directly or indirectly to human exposures, an issue that is addressed in

Chapter 4. The exposure pathways resulting from human intrusion are not depicted on Figure 3.2.

Given that the important pathways that could lead from the repository to human exposure have been identified, in a general sense, the technical feasibility of developing performance assessment calculations to evaluate compliance with a risk standard for Yucca Mountain depends on the feasibility of modeling the relevant processes that lead to transport, accompanied by dilution or concentration along these pathways. It also depends on the feasibility of quantifying the probabilities associated with any processes that cannot be predicted in a purely deterministic fashion. As discussed in the preceding section of this chapter, uncertainties can be associated with conceptual, mathematical, and numerical models of processes. Uncertainties in parameters and in boundary conditions also necessitate a probabilistic treatment. In some cases, reasonable estimates can be made only of the bounds of probability of occurrence. In other cases, only the consequences of a process can be estimated or bounded in a quantitative manner. Some uncertainties could be such that the probabilities of occurrence or consequences are not quantifiable. The time frame over which models must be applied also influences the level of uncertainty.

By soliciting the opinions of knowledgeable scientists at our meetings, through review of solicited and unsolicited written contributions, and drawing on the available literature and our own expertise, we attempted to assess the types, magnitudes, and time-dependencies of uncertainties associated with (1) transport from a Yucca Mountain repository through the various reservoirs shown in Figure 3.2; (2) the effects of potential natural and human modifiers to repository performance; and (3) the exposure pathways through the biosphere. Part II below summarizes our conclusions regarding the feasibility of including the geologic and physical factors in a quantitative performance assessment. The inclusion of human behavior and other exposure factors are discussed in Part III of this chapter. In carrying out this evaluation of the feasibility of quantitative performance assessment of a repository, we did not attempt to evaluate the performance of a repository at Yucca Mountain.

PART II: EARTH SCIENCE AND ENGINEERING FACTORS IN PERFORMANCE ASSESSMENT AT YUCCA MOUNTAIN

Our conclusions about the feasibility of using quantitative performance assessment at Yucca Mountain rely on a systematic analysis of the application of the methodology to this specific site.

Transport Among Reservoirs

For the processes leading to transport among the reservoirs identified in Figure 3.2, we conclude that the processes are sufficiently quantifiable and that the uncertainties are sufficiently boundable that they can be included in performance assessments that extend over time frames corresponding to those over which the geologic system is relatively stable or varies in a boundable manner. The geologic record suggests that this time frame is on the order of about 10^6 years. Some of the important considerations for the reservoirs and associated transport processes are summarized below.

Release from the waste form

Calculations of release rates from the waste-packages require information on waste composition, waste-package properties, and the thermal, chemical and hydrologic processes that can lead to deterioration and failure of canisters. If the two major waste sources are spent fuel from light-water power reactors and borosilicate glass containing defense waste, it would appear that the necessary engineering parameters relating to the waste form could be reasonably well specified. Specification of waste properties will be more complicated if large amounts of waste with more heterogeneous properties are included. Time-dependent temperatures, porosity and humidity in the vicinity of the canisters, required for estimates of waste package failure, are currently being calculated using models of two-phase convective flow in the near-field unsaturated zone induced by thermal loading of the repository. Detailed estimates of time for canister failure are less important for much longer-term estimates of individual dose or risk.

Simplifications in predictive models of waste package failure, as currently employed for the Yucca Mountain project, do not account for diffusive impedance of penetrations.[1] Recent studies show that diffusive impedance from penetrations results in release rates being insensitive to water flow rates near the waste package, at least for some period of time. These results suggest that uncertainties in release-rates resulting from uncertainties in water flow rate might be small even for sites with pore water velocities much greater than those expected at Yucca Mountain. Further refinements in release-rate predictions can be made if the time-dependent failure characteristics of waste containers and of Zircaloy cladding on spent fuel can be estimated.

Transport from canisters to the near-field unsaturated zone

Inflow of air through failed canisters and oxidation of waste prior to infiltration of water can affect the time-dependent release rates of gaseous radionuclides as well as the later release of dissolved radionuclides. This process would probably affect estimates of 10,000-year cumulative releases more than estimates of longer-term doses and risks. Once a waste canister has been penetrated, release of soluble radionuclides will be affected by the size of canister penetration, diffusion coefficients, rates of oxidative alteration of spent fuel and hydration or alteration of borosilicate glass defense waste. Water content and porosity of the surrounding rock or backfill can affect rates of air and water ingress that can contribute to oxidation and hydration of the waste. Release rates of solubility-limited elements such as thorium, uranium, neptunium, plutonium and americium depend on the chemical environment within the waste canister as well as on the physical properties that affect more soluble species. A conservative release model for low solubility species can be strongly affected by the local flow rate of unsaturated water because all water flowing past the waste is assumed to become saturated with solubility-limited species. This leads to predicted cumulative releases being sensitive to water flow rate in the vicinity of the waste. In contrast,

---

[1] Canisters are likely to fail initiallly at small local openings through which water might enter, but out of which the diffusion of dissolved wastes will be slow until the canister is grossly breached.

the moist-continuum model used in recent Yucca Mountain assessments predicts that release rates will be dominated by molecular diffusion, with little dependence on water flow rate at the low pore velocities in the vicinity of the waste packages.

Colloid formation has the potential to increase bulk concentration of radionuclides in water adjacent to the waste-form surface. The Yucca Mountain project has not yet implemented analyses of colloid formation and transport affected by filtration and interactions with radioactive solutes. Sorption in backfill and rock surrounding the waste package could substantially retard diffusion of radionuclides away from the waste canister. Analysis of diffusion for unsaturated zone sites such as Yucca Mountain, with or without sorptive retardation, is currently limited by knowledge of the effective diffusion coefficients in unsaturated backfill and tuff. In principle, however, these processes are amenable to the type of quantitative modeling required for performance assessment.

Gas phase transport from the unsaturated zone to the atmosphere above Yucca Mountain

Some radionuclides released from the waste forms, of which carbon-14 (^{14}C) is probably the most important, can be mobile in the gas phase of the unsaturated zone. Gas phase transport can lead directly to releases to the biosphere when the gas flows out of the mountain into the near-surface atmosphere. Diffusive and convective transport in the gas phase are both likely to reduce concentrations within the unsaturated zone as contaminated and uncontaminated air mix during transport through the mountain. Further mixing and concentration reductions will occur once the air is released from the unsaturated zone to the atmosphere. These mixing processes can have significant effects on individual doses and risk since they will control the concentrations to which humans are exposed, largely through the consumption of ^{14}C in plants. Concentrations will be reduced by radioactive decay only if time elapsed from emplacement in the repository to release at the land surface is long compared to the half-life of the radionuclide.

The major sources of uncertainty in the calculation of local exposures from gaseous releases of $^{14}CO_2$ from Yucca Mountain are canister life, time distribution of canister failure, the fraction of ^{14}C

initially released when a canister and the fuel rods it contains are breached, the release rate of the ^{14}C contained in the ceramic matrix, and the dispersion of the ^{14}C when it is released into the air at Yucca Mountain. The mechanisms of gas phase transport are fairly well understood, and the available evidence suggests that travel times from the repository, once ^{14}C is released from the waste canisters, are comparatively short.

Atmospheric circulation leading to dispersal of gaseous radionuclides in the world atmosphere

An estimate of radionuclide concentrations in air resulting from releases at Yucca Mountain can be made assuming that radionuclides are distributed uniformly through the world atmosphere. However, this case provides a global average estimate for individual exposures, and would not indicate whether higher exposures might occur at specific locations. More sophisticated models are also available and have been applied to Yucca Mountain release scenarios (see calculations presented in Chapter 2). Calculations of this type have been used to assess potential population doses of ^{14}C and compare these with a negligible incremental dose limit. Such calculations would be directly applicable to quantitative assessments of compliance with a population-risk standard. In addition to the analysis of local individual exposures that might result from gaseous releases at Yucca Mountain, there have been numerous studies of the global effects of such releases (Nysaard et al., 1993). For ^{14}C, the dominant pathway is through the uptake of $^{14}C0_2$ by plants and the ingestion of those plants by humans. If the level of health risk is as these studies suggest, the average global exposures that would result would be classified as negligible individual doses, as described in Chapter 2. The standard that we recommend would include local risks from ^{14}C in its analysis. If those risks were found to be significant, they would be included against the risk limit we propose.

Aqueous phase transport from the unsaturated zone to the water table

Mechanisms of aqueous phase transport of dissolved radionuclides in the unsaturated zone at Yucca Mountain are less well understood than those of gas phase transport. The porous flow and dual porosity models employed to date in performance assessment exercises for Yucca Mountain have been criticized (L. Lehman, L. Lehman and Associates, personal communication, Dec. 16, 1993) for not incorporating adequate representations of the controlling features, particularly episodic flow through fractures. This is an example of uncertainty in the underlying conceptual model. DOE and its contractors recognize some of the limitations of the current models and are evaluating alternative unsaturated zone flow and transport codes as part of site characterization activities (Reeves et al., 1994). According to Reeves et al., none of the existing codes identified has the adequate capabilities to simulate the nonequilibrium fracture-matrix flow that might arise during unsteady infiltration in the unsaturated zone at Yucca Mountain. However, we have been presented with results from detailed analyses by Nitao et al., (1993) that do consider episodic nonequilibrium fracture-matrix flow.

Uncertainties in unsaturated zone travel time estimates are most significant for standards that are applicable over a limited time frame. For an individual-risk standard, the significance of these travel time estimates is that they determine the time available for radioactive decay. Long travel times would allow for significant decay and, as a result of the decay, reduction in radionuclide fluxes to the water table. Unsaturated zone travel times for some radionuclides can be increased by sorptive retardation. Uncertainties in retardation estimates stem from the limited amount of data on sorption isotherms of the radionuclides with the various rock units at Yucca Mountain. This uncertainty can be reduced through additional laboratory studies to measure these isotherms. If unsaturated zone transport occurs primarily by episodic, rapid flow through fractures, it is possible that sorption isotherms might overestimate sorptive retardation, at least during the period of fracture flow. Solution phase complexation and sorption to mobile colloids would also serve to limit retardation. Conservative bounding calculations in such cases would be those that consider the radionuclide to behave as a nonsorbing solute.

Although considerable uncertainty currently exists regarding the mechanism and rates of aqueous phase transport in the unsaturated zone, these uncertainties do not preclude incorporation of this transport in a quantitative performance assessment. Site characterization activities currently underway are designed to elucidate the processes and provide improved estimates of the relevant parameters. Even if these efforts are of limited success in reducing uncertainties, bounding estimates can be incorporated into a performance assessment designed to evaluate compliance with an individual risk standard.

Saturated zone transport from the aquifer beneath the repository to other locations from which water may be extracted by humans or ultimately reach the surface in a regional discharge area

The time at which inhabitants downgradient from a Yucca Mountain repository could be exposed to radionuclides depends on the rates of advective transport in the saturated zone and on modifications to that rate resulting from geochemical processes such as sorption. Rates of advective transport in the saturated zone can be estimated using existing models that require quantification of the hydraulic properties of the rock and of the hydraulic gradient. Modification in transport rates by geochemical processes depends on the rate and extent of chemical interactions between the dissolved radionuclides and the aquifer solids. Geochemical processes can also modify concentrations of radionuclides in ground water. Concentrations can also be modified by radioactive decay, by diffusion, and by dispersive mixing of contaminated and uncontaminated water. Thermal gradients induced by the repository could generate additional convective mixing that would reduce peak concentrations beneath the repository.

The important processes of saturated zone transport are understood at a conceptual level, and mathematical models are available to represent these processes to some extent. Because of the fractured nature of the tuff aquifer below Yucca Mountain, some uncertainty exists regarding the appropriate mathematical and numerical models required to simulate advective transport. This issue can be addressed through the site-characterization activities and through sensitivity modeling. Major uncertainties regarding the values of hydraulic and geochemical parameters required as input to these models are likely to remain even at the end of extensive site characterization due to the inherently heterogeneous nature of the aquifer. However, even with residual uncertainties, it should be possible to generate quantitative (possibly bounding) estimates of radionuclide travel times and spatial distributions and concentrations of plumes accessible to a potential critical group.

Gradual and Episodic Natural Modifiers

Several gradual and episodic natural processes or events have the potential to modify the properties of the reservoirs and the processes by which radionuclides are transported among them. We conclude that the probabilities and consequences of modifications generated by climate change, seismic activity, and volcanic eruptions at Yucca Mountain are sufficiently boundable so that these factors can be included in performance assessments that extend over periods on the order of about 10^6 years. Each of these three possible modifiers of repository performance is discussed in more detail below.

Climate change

At present the earth is in an interglacial phase. Our knowledge of past climate transitions indicates that a transition to a glacial climate during the next few hundred years is highly unlikely but not impossible. Such a transition during the next 10,000 years is probable, but not assured. Over a million-year time scale, however, the global climate regime is virtually certain to pass through several glacial-interglacial cycles, with the majority of the time probably spent in the glacial state. Given that a deep geologic repository is relatively shielded from the large changes in surface conditions, there are three main potential effects of climate change on repository performance. The first of these is that increases in erosion might significantly decrease the burial depth of the repository. Site-specific studies of erosion rates at Yucca Mountain (DOE 1993b) indicate that an increase in erosion to the extent necessary to expose the repository (even over a million-year time scale) is extremely unlikely.

Change to a cooler, wetter climate at Yucca Mountain would likely result in greater fluxes of water through the unsaturated zone, which could affect rates of radionuclide release from waste-forms and transport to the water table. Little effort has been put into quantifying the magnitude of this response, but a doubling of the effective wetness, defined as the ratio of precipitation to potential evapotranspiration, might cause a significant increase in recharge. An increase in recharge could raise the water table, increasing saturated zone fluxes. There is a reasonable data base from which to infer past changes in the water table at Yucca Mountain. Although past increases under wetter climates are evidenced, a water-table rise to the point that the repository would be flooded appears unlikely (Winograd and Szabo, 1988; NRC, 1992; Szabo et al., 1994). Additional site characterization activities and studies of infiltration at Yucca Mountain should help improve estimates of the bounds of potential hydrologic responses to climate change. It should also be noted that the subsurface location of the repository would provide a temporal filter for climate change effects on hydrologic processes. The time required for unsaturated-zone flux changes to propagate down to the repository and then to the water table is probably in the range of hundreds to thousands of years. The time required for saturated flow-system responses is probably even longer. For this reason, climate changes on the time scale of hundreds of years would probably have little if any effect on repository performance, and the effects of climate changes on the deep hydrogeology can be assessed over much longer time scales.

The third type of change that might result from climate change is a shift in the distribution and activities of human populations. In the vicinity of Yucca Mountain, a wetter, cooler climate would provide a more hospitable environment and could result in population increases. This could change the composition of the critical group by exposing more people to potential risks from the repository. However, even at the present time, the available ground-water supply could sustain a substantially larger population than that presently in the area. Thus, there is no simple relation between future climatic conditions and future population. This unpredictability of human behavior is common to the issue of estimating pathways through the biosphere and will be addressed later in Part III.

Seismicity

Seismic displacement along faults is one type of episodic event that must be considered in estimating the long-term safety of a repository at Yucca Mountain. The adverse effects of seismicity can be assessed in terms of canister failure or an increase in fluid conductivity in the saturated or unsaturated zone. Yucca Mountain is within a region of Quaternary (from 2 million years ago to the present) seismic activity, of which the Little Skull Mountain earthquake of June 29, 1993, with a Richter magnitude 5.6, is the most notable recent example. Measured slip rates on faults in the region vary from approximately 0.001 mm/yr to 0.02 mm/yr with recurrence intervals of 20,000 to 100,000 years (Whitney, 1994). Also, according to Whitney, no significant faults, that is, faults with more than 5 cm displacement over the last 100,000 years have been found at the proposed repository site. Seismicity is an episodic process, appearing to be essentially a fractal activity involving frequent releases of small amounts of strain energy and progressively less frequent releases of larger amounts of energy. It is possible through careful examination of the geologic record to establish a chronological history of the activity over millions of years. Estimates of activity over similar periods into the future can be made by extrapolation from the past activity.

Seismic effects are important both during the repository operational or pre-closure phase and the post-closure phase. The effect of a seismic event on underground excavations, such as repositories is usually less severe than the effects on a surface facility. Numerical models are available to assess the effect of seismicity on displacements along fractures and faults in rock. It would appear that, with good engineering, the probability of adverse effects on repository isolation capabilities due to seismic loading at Yucca Mountain could be reduced sufficiently to result in boundable and probably very low risk.

Specifically, with respect to the effects of seismicity on canisters, the rock mass at Yucca Mountain is extensively fractured so the future seismic displacements are likely to occur along existing fractures rather than on new ones. Risks could be further reduced through the practice of "fault avoidance," whereby no canisters would be placed within or immediately adjacent to a known underground fault (which should be readily apparent during excavation of repository drifts and canister emplacement holes). Similarly, in-drift placement of canisters surrounded by a buffer backfill, such as bentonite-sand could essentially isolate canisters from the effects of seismicity.

With respect to the effects of seismicity on the hydrologic regime, the possibility of adverse effects due to displacements along existing fractures cannot be overlooked. It would seem that the hydrologic regime has been conditioned by many similar seismic events over geologic time. In consequence, such displacements have an equal probability of favorably changing the hydrologic regime, so that the effect of seismicity on the hydrologic regime could probably be bounded.

Studies have been made of the possibility that a seismic event could produce transient changes in the water table at Yucca Mountain sufficient to bring ground water through the repository to the surface (NRC, 1992). Results indicate a probable maximum transient rise on the order of 20 m or less. In summary, although the timing of seismic events is unpredictable, the consequences of these events are boundable for the purpose of assessing repository performance.

Volcanism

A volcanic intrusion into the proposed repository could be catastrophic, releasing a major part of the repository inventory directly into the biosphere. However, the overall risk might be very low, because it is also a very unlikely event. Like seismicity, volcanism is episodic. The two phenomena could also be linked, in that some seismic activity can be triggered during periods of volcanic activity. Unlike seismicity, volcanism in the Yucca Mountain region involves intermittent concentrated activity separated by long repose periods. Even so, like seismicity, estimates of future volcanic activity can be based on analysis of the geologic record, with the assumption that the same pattern of events will hold in the future.

The risk from volcanism at Yucca Mountain is being examined using a probabilistic approach. According to Crowe et al. (1994), current studies are designed to establish three components of an overall probability of magmatic disruption of a repository:

1. Future recurrence rate of volcanic events, such as volcanic centers or volcanic clusters;
2. The probability that a future event will intersect a specified area, such as the repository or a controlled area beyond the repository;
3. The probability that an event occurring within the specified area will release radionuclides into the biosphere.

The probability of occurrence of the second component depends upon the probability of the first component, and the overall probability of radionuclide release due to volcanism in the Yucca Mountain region depends on the combined probability of all three components. Emphasis is being given to estimating the combination of the first and second components to determine the combined probability that a future event will intersect a specified area. This analysis is based on extrapolations into the future of volcanic activity from the historic record, and on assumptions about the spatial distribution of future volcanic eruptions in the Yucca Mountain region. Crowe suggests that a probability of 10^{-8}/yr, which is a 1 in 10,000 possibility of a disruption over 10,000 years or 1 in 1,000 possibility in 100,000 years) or less might be sufficiently low to constitute a negligible risk. If the combined probability of the first two components can be shown to be below this level, then it might not be necessary to consider the third component.

Efforts are underway to refine the intrusion distribution models by incorporating geologic structure constraints. It is noted, for example, that the volcanic eruptions in Crater Flat appear to be aligned in the northeast direction of the extensional faulting (across the Yucca Mountain site). If this constraint is confirmed and included in the distribution, the probability of a future event intersecting the repository site might fall below 10^{-8} per year.

While acknowledging the complexity of estimating the release of radionuclides to the biosphere, it seems possible, given the knowledge of material ejected from various types of volcanic eruptions and study of the cinder cones in the region, to develop reasonable estimates of the health consequences from radionuclides released by a volcanic eruption through a repository at Yucca Mountain. Thus, it is believed that the radiological health risk from volcanism can and should be subject to the overall health risk standard to be required for a repository at Yucca Mountain.

PART III: EXPOSURE SCENARIOS IN PERFORMANCE ASSESSMENT

As noted above, we believe that it is feasible to calculate, to within reasonable limits of certainty, potential, defined as possible but not necessarily probable concentrations of radionuclides in ground water and air at different locations and times in the future. To proceed from the calculation of radionuclide concentrations to calculations of risks that would result from a repository, many additional factors or assumptions about the nature of the human society at or near the repository site must be considered. These factors must be included in an exposure scenario that specifies the pathways by which persons are exposed to radionuclides released from the repository.

As we note in Chapter 4 with regard to the feasibility of making projections of future human intrusion into a repository, based on our review of the literature we believe that no scientific basis exists to make projections of the nature of future human societies to within reasonable limits of certainty. Therefore, unlike our conclusion about the earth science and geologic engineering factors described in Part II of this chapter, we believe that it is not possible to predict on the basis of scientific analyses the societal factors that must be specified in a far-future exposure scenario. There are an unlimited number of possible human futures, some of which would involve risks from a repository and others that would not.

Although the nature of future societies cannot be predicted, it is possible, at least conceptually, to consider several characteristics of future society that would indicate whether a repository is likely to pose a risk to people. A repository would be unlikely to pose significant risks to future societies: if the area near the repository were not occupied, if future societies do not use ground water from the contaminated region, or if future societies routinely monitor ground-water quality and either treat or avoid use of contaminated sources. Conversely, exposures would result if water wells were drilled into the contaminated areas and the water consumed by people or used to irrigate crops. As far as we are able to determine, there is no sound basis for quantifying the likelihood of future scenarios in which exposures do or do not occur; about all that can be said is that both are possible.

It is our view, however, that once exposure scenarios have been adopted, performance assessment calculations can be carried out for the specified scenarios with a degree of uncertainty comparable to the uncertainty associated with geologic processes and engineered systems. The more difficult task is the specification of reasonable scenarios for evaluation. Any particular scenario about the future of human society near Yucca Mountain that might be adopted for purposes of calculation is likely to be arbitrary, and should not be interpreted as reflecting conditions that eventually will occur. Although we recognize the burden on regulators to avoid regulations that are arbitrary, we know of no scientific method for identifying these scenarios.

Selection of Exposure Scenarios for Performance Assessment Calculations

Any approach to assessing compliance with the standard must make assumptions about the nature of the human activities and lifestyles that provide pathways for exposure. For example, people could drink water containing radionuclides, irrigate crops with the water, eat these crops, and bathe in the water. Quantification of the doses received from the various pathways requires detailed data on these pathways. For the example above, the average amount of water ingested per day (not including other beverages constituted with uncontaminated water) should be known, as should the type of crops grown, the amount eaten, and the frequency of bathing. The set of circumstances that affects the dose received, such as where people live, what they eat and drink, and other lifestyle characteristics including the state of agricultural technology, are part of what we refer to as the exposure scenario.

Unfortunately, many human behavior factors important to assessing repository performance vary over periods that are short in comparison with those that should be considered for a repository. The past several centuries (or even decades) have seen radical changes in human technology and behavior, many or most of which were not reasonably predictable. For example, within the past one hundred years, our society has evolved from one in which drilling and pumping technology did not exist for production of water from the depths of ground water at Yucca Mountain to a level of technology where such production is feasible. Within this same time period, we have seen U.S. demographic patterns shift from a time where a majority of U.S. residents were engaged in farming and grew their own food to the present day in which only a few percent of the work force is employed in farming, and in which most people's diet includes food produced outside their local area.

Given this potential for rapid change, it is unknowable what patterns of human activity might exist 10,000 or 100,000 years from now. Indeed, the period during which repository performance might be relevant, on the order of a million years, is sufficiently long that any number of different societies might reside near the repository site. Several glacial periods probably will have occurred, making estimates of human society even more difficult. Given the unknowable nature of the state of future human societies, it is tempting to seek to avoid the use of such assumptions in performance assessment calculations. In our view, however, it is not possible for a reasonable standard for the protection of human health to avoid use of some specified assumptions about future populations, patterns, and lifestyles around a proposed repository site. Even regulatory standards stated in terms of geologic and engineering factors are not independent of assumptions about future exposure scenarios. For example, the containment requirements of 40 CFR 191 were apparently developed based on consideration of a global release scenario in which average doses to large populations were considered.

The problem is how to pick an exposure scenario to be used for compliance assessment purposes. Given the lack of a scientific basis for doing so, we believe that it is appropriate for the regulator to make this policy decision. One specific recommendation we make is to avoid placing the burden of postulating and defending assumptions about exposure scenarios on the applicant for a license. The regulator appears to be better situated than the applicant to carry the responsibility because of the perception that any future scenario developed by the applicant could have been chosen to give the desired outcome. On the other hand, the results of calculations from a scenario specified by the regulator in an open process designed to consider the views of all the interested parties might be seen as a fair test of the suitability of a site and design.

In addition, we recommend against an approach under which a large number of future scenarios are specified for compliance assessment, since such an approach could be seen as putting both the regulator and the applicant in the indefensible position of claiming to have considered a sufficient number of scenarios and that all reasonable future situations are represented in the analysis. The purpose of making exposure scenario assumptions is not to identify possible futures, but to provide a framework for the analysis and evaluation of repository performance for the protection of public health.[2]

---

[2] Another argument for using a large number of scenarios is that iterative analysis of repository performance will lead to the most cost-effective repository design. This might be true, but we believe that the regulator must in the end assess compliance with a single level of protection as defined in the standard. Therefore, one (or at most a few) exposure scenarios must be specified for compliance assessment purposes.

Specification of the exposure scenario assumptions to be used in performance assessment at Yucca Mountain will greatly influence whether the site and design can comply or not. The selection of exposure scenarios is perhaps the most challenging and contentious aspect of risk and compliance assessment. For example, EPA guidlines for exposure assessment reflect a philosophical disagreement over the question of when and how to depart from the theoretical upper bound estimate of exposure and to employ probabilistic techniques (Federal Register 57 [May 29, 1992]: 22888-22938). These questions, which are at the interface between science and policy judgment, are also addressed in *Science and Judgment in Risk Assessment* (NRC, 1994). For these reasons, we strongly recommend that the decision be made through a public rulemaking process. This process will provide a more complete analysis of the advantages and disadvantages of alternative scenarios than we have been able to perform, and do so with the benefit of full public participation.[3]

As with other aspects of defining the standards and demonstrating compliance that involve scientific knowledge but must ultimately rest on policy judgments, we considered what to suggest to EPA as a useful starting point for rulemaking on exposure scenarios. Reflecting the disagreement inherent in the literature, we have not reached complete consensus on this question.

We do agree, however, that the exposure scenario used to test compliance should not be based on an individual defined by unreasonable assumptions regarding habits and sensitivities affecting risk. It is essential that the exposure scenario that is ultimately selected be consistent with the critical-group concept that we advanced in Chapter 2. The purpose of using a critical group is to avoid using the standard to protect a person with unusual habits or sensitivities. The critical-group approach does this by using the average risk in the group for testing compliance. To ensure that this average risk nevertheless affords a high level of protection to most persons, the group must contain the persons at highest risk within the group and must be homogeneous in risk. An exposure scenario selected for

---

[3] This rulemaking need not be done before the promulgation of an individual-risk standard that we recommended in Chapter 2. Indeed, we would not want the selection of that standard to be colored by foreknowledge of the assumptions incorporated in the exposure scenario.

compliance assessment should produce a critical group with these characteristics.

Additionally, we note that the ICRP (1985a) recommends that the critical group be defined using present knowledge[4] and cautious, but reasonable, assumptions. Although this guidance was originally intended for the regulation of dose limits, we believe that it is generally appropriate in applying the critical-group concept to risk, as we have recommended. EPA should rely on this guidance when choosing the assumptions for the exposure scenario to be used for performance assessment.

Finally, we have considered the design of an exposure scenario that EPA might propose when it initiates the rulemaking process. We have considered two illustrative approaches for this purpose. We describe the two approaches in Appendixes C and D, and summarize their important characteristics below.

A substantial majority of the committee considers that the approach outlined in Appendix C is more clearly consistent with the foregoing criteria for selecting an exposure scenario than is the alternative in Appendix D, and therefore believes that EPA should propose an approach along the lines of Appendix C. Of course, other methods might also meet these criteria, and some of the methods might be less complex than the method illustrated in Appendix C.

Although the following discussion highlights differences between the two approaches, we wish to stress that the approaches are similar in many ways.

The approach in Appendix C makes use of information that can be collected on the factors that influence human behavior in the present. Assumptions about factors such as the source of food would be based on the source of food for today's population near the repository site. The Appendix C approach bases the exposure scenario on a population distribution derived from observed statistical associations between environmental parameters and the population distribution of actual population groups. For example, such parameters could include depth to

---

[4] We understand "present knowledge" to mean any knowledge that is available today, and so should be read as an injunction against making assumptions about knowledge that might exist in the future. For example, assuming that future societies will have found a cure or prevention for cancer would not be present-day knowledge.

water, soil type and depth, land slope, and growing season. This approach uses statistical techniques to compute a critical group for each of a large number of simulations of the contaminated ground-water plume and then averages over these calculations to identify the average critical group for compliance purposes.

Important characteristics of this approach include the following. First, it extends the probabilistic methods that have been applied to simulations of physical processes (such as transport of ground-water contaminants) to analysis of the factors affecting exposure. Second, although mathematically complex, the model is based on currently observable data and does not require assumptions regarding specific values of parameters, only ranges within which the parameters might fall. Third, the degree to which conservatism is incorporated is determined not only by the analyst in selecting the ranges of parameters that describe farming lifestyles but also by the regulator when the standard is set. Fourth, it requires that the probability that persons occupy specific parcels of land for farming be determined statistically by the relevant characteristics of the land, ground water, and technology that influence farming, avoiding the potential that the standard could be influenced by a situation in which the maximum dose occurred at a place that was uninhabitable or otherwise unsuitable for farming.

The approach in Appendix D specifies *a priori* one or more subsistence farmers as the critical group and makes assumptions designed to define the farmer at maximum risk to be included in the critical group. The subsistence farmer would be a person with eating habits and with response to doses of radiation that are normal for present-day humans. All food eaten over the lifetime of the subsistence farmer would be grown with water drawn from an underground aquifer contaminated with radioactivity from the repository. The water would be withdrawn at a location outside the footprint of the repository and near that maximum potential concentration of the most critical radioactive contaminant in the ground water so that the scenario describes the maximum dose and risk. All of the farmer's drinking water would come from that same source. For compliance assessment purposes, it is assumed that the homogeneity criterion (see the definition of critical group in Chapter 2) applies and that the risk to the average member of the critical group is about one-third that of the subsistence farmer.

The important features of the subsistence-farmer model include the following. First, it has been used extensively in radioactive waste management programs in the United States and other countries, so a body of experience with it exists on which to draw. Second, it is straightforward and relatively simple to understand and calculate. Third, while it incorporates a series of assumptions about the lifestyle of the hypothetical farmer, any degree of conservatism can be built into the model by choices among alternative assumptions, which can be based on current conditions in the Amorgosa Valley; these assumptions need not be constrained by the characteristics of the current population of the region. Fourth, it makes the most conservative assumption that wherever and whenever the maximum concentration of radionuclides occurs in a ground water plume accessible from the surface, a farmer will be there to access it.

These approaches have many elements in common. Most important, both rely on probabilistic methods of estimating the distribution of radionuclides in the environment. Both also incorporate knowledge of the natural geologic features of the environment that influence the potential for exposure and both are intended to incorporate cautious, but reasonable, assumptions about lifestyles of the affected populations that the EPA might propose in a rulemaking. For example, both assume eating habits and response to radiation doses that are normal for present-day humans.

Despite these similarities between the approaches, two major issues that differentiate them have emerged from our consideration. These issues are summarized below:

- Assumptions about the location and lifestyle of persons who might be exposed to radionuclides released from the repository are crucially important because they affect the identification of the person at highest risk that must be contained in the critical group. The two approaches differ in their treatment of these assumptions. For example, the approach in Appendix D specifies *a priori* that a person will be present at the time and place of highest nuclide concentrations in ground water and will have such habits as to be exposed to the highest concentration of radiation in the environment. This person is assumed to define the upper limit of risk in the critical group. Appendix C treats the distribution of potential farmers probabilistically based on current technical understanding of farming in the region. Because the person at highest risk might not be the same under the two approaches, the critical group selected for compliance assessment could be different.

- The second difference involves the method of calculating the average risk of the members of the critical group. Appendix C uses detailed statistical analysis to define the critical group. Specifically, it identifies a "critical subgroup" for each of a large number of Monte Carlo realizations of the contamination plume. The critical group risk is determined by averaging over the average risks to each of these subgroups. In contrast, the Appendix D approach approximates the average critical group risk at about one-third of the risk faced by the person at highest risk, since the requirement that the critical group be homogeneous in risk implies that the overall range of risks in the critical group be limited to about a factor of ten. If the distribution of risk among members of the critical group is not relatively uniform, these approaches could produce different averages.

As noted earlier, we agree that unrealistic assumptions are inappropriate. Our divergence of view is on the extent to which the alternative sets of assumptions embodied in Appendixes C and D are cautious, but reasonable. The approach of Appendix C has the advantages of explicitly accounting for how the physical characteristics of the site might influence population distribution and of identifying the makeup of the critical group probabilistically. Most of the committee regard these as desirable features of exposure scenarios that are intended to be consistent with the critical-group concept. We emphasize, however, that specification of exposure-scenario assumptions is a matter for policy decision.

Exclusion Zone

The original standard, 40 CFR 191, contained a provision for an exclusion zone in the immediate vicinity of the repository. The purpose was to provide a boundary for calculating releases.

In light of our conclusion in Chapter 4 that there is no scientific basis for assuming that institutional controls can be maintained for more than a few centuries, we also conclude that there is no scientific basis for assuming that human activity can be prevented from occurring in an exclusion zone or that defining such a zone will provide protection to future generations from exposures in the vicinity of the repository.

The question remains whether an exclusion zone serves a useful purpose for compliance assessment. In our analysis, we have assumed that some human activities, such as drilling into or through the repository, should be treated as special cases of human intrusion (see Chapter 4). If, as we recommend, human intrusion is treated separately from the performance of an undisturbed repository, it is reasonable in our view to define a region in which human activities are to be regarded as intrusion and to exclude that region from calculation of the undisturbed repository performance. For example, if we assume that all drilling for water wells is vertical, the area directly above the repository plan (or footprint) would be considered an exclusion zone for purpose of calculating compliance with that part of the standard that applies to undisturbed performance. Drilling in that zone would be a case of human intrusion.

Beyond the repository footprint, however, there seems to be no practical purpose for defining a larger exclusion zone for the form of the standard we recommend. Without either a release limit or a time limit for the standard for undisturbed performance, an arbitrary boundary serves no purpose. In the approach we recommend, an objective of performance assessment calculations is to determine the time in the future when risks from exposure to radionuclides released from the repository are greatest and to base the regulatory judgment about compliance on a comparison of the risks at that time to the standard. Furthermore, neither of the alternatives for treating the critical group requires an exclusion zone larger than the repository footprint.

4

HUMAN INTRUSION AND INSTITUTIONAL CONTROLS

INTRODUCTION

In Section 801(a)(2) of the Energy Policy Act of 1992, Congress asked three specific questions. The first question, about the use of individual dose as a criterion for protecting the public, was addressed in Chapter 2. The second and third questions concern the potential that at some future time people might intrude into the repository, thereby defeating its geologic and engineered barriers. We were asked to examine the scientific basis for predicting human intrusion and the potential for protecting against it, specifically:

> Question 2. Whether it is reasonable to assume that a system for post-closure oversight of the repository can be developed, based on active institutional controls, that will prevent an unreasonable risk of breaching the repository's engineered barriers or increasing the exposure of individual members of the public to radiation beyond allowable limits.
>
> Question 3. Whether it is reasonable to make scientifically supportable predictions of the probability that a repository's engineered or geologic barriers will be breached as a result of human intrusion over a period of 10,000 years.

Briefly, we conclude that the answer to both questions is "no" for the reasons outlined below.

Human activity that penetrates the repository, such as by drilling into it from the surface, can cause or accelerate the release of radionuclides. Waste material might be brought to the surface and expose the intruder to high radiation doses, or the material might disperse into the biosphere. Even if this does not occur, a borehole could go through the repository and open a pathway by which radionuclides more readily reach the ground water.

Over the years, DOE has developed a considerable literature on human intrusion and on active and passive controls to prevent it (von Winterfeldt, 1994). For example, some studies have examined resource potential and historical exploration activity and used current understanding and rates of drilling to project future activity. Other studies have detailed examples of monuments and inscriptions that have survived from long ago. Still others have speculated on the characteristics of signs and markers that might improve their long-term effectiveness at delivering a message to future generations. Based on our understanding of this literature, however, we conclude that there is no technical basis for predicting either the nature or the frequency of occurrence of intrusions.

For some initial period, human intrusion could be managed through active or passive controls. As long as they are in place, active institutional controls such as guards could prevent intruders from coming near the repository. We conclude, however, that there is no scientific basis for making projections over the long term of either the social, institutional, or technological status of future societies. Relying on active controls implies requiring future generations to dedicate resources to the effort. There is, however, no scientific basis from which to project the durability of governmental institutions over the period of interest, which exceeds that of all recorded human history. On this time scale, human institutions have come and gone. We might expect some degree of continuity of institutions, and hence of the potential for active institutional controls, into the future, but there is no basis in experience for such an assumption beyond a time scale of centuries.

Similarly, there is no scientific basis for assuming the long-term effectiveness of active institutional controls to protect against human intrusion. Although it may be reasonable to assume that a system of post-closure oversight can be developed and relied on for some initial period of time, there is no defensible basis for assuming that such a system can be relied on for times far into the future. Between these limits, the ability to rely on such active institutional systems presumably diminishes in a way that is intrinsically unknowable. We have seen no evidence to support a claim to the contrary. People might disagree, of course, on their predictions for how long into the future active institutional controls might survive and remain effective.

The situation is not qualitatively different for passive institutional controls. As long as they are recognized and heeded, passive controls such as markers, barriers, and archival records could serve to warn potential intruders away. Passive controls, too, may be of limited duration, requiring future generations to renew them. While many historical markers, monuments, and records have survived for long periods of time, up to thousands of years, most presumably have not. Those records that have survived might not represent records for which the local social knowledge was continuous. We cannot know those that did not survive to our time. Further, languages have changed over periods of centuries so that old documents and inscriptions might be difficult for any but scholars to interpret. Even though technologies for making markers and monuments will improve and even though modern global telecommunications might slow the rate of change of languages, the time span of concern for a high-level waste repository far exceeds experience, so there is no technical basis for making forecasts about the reliability of such passive institutional controls.

Just as there is no basis for assuming the effectiveness of either active or passive institutional controls to reduce the risk of human intrusion, we also conclude that there is no scientific basis for estimating the probability of intrusion at far-future times. Several types of intrusion can be considered: inadvertent intrusion into the repository in the process of exploring for or producing other resources in the vicinity, intrusion driven by curiosity about the markers and what might lie below them, or intentional intrusion for malicious purposes or to recover the repository contents. (The malicious intrusion might be by a hostile nation or subnational group assuming a societal or institutional presence.) In our view, there is simply no scientific basis for estimating the probability of inadvertent, willful, or malicious human action.

Estimating the probability of inadvertent intrusion as a consequence of exploration or production of resources might seem more plausible than for the cases of willful or malicious intrusion. Doing so, however, requires knowledge of which materials at or near the site will be regarded as resources in the future and the technologies that will exist for exploration and production. We cannot predict future economic conditions that help to define what is a valuable resource nor can we forecast future exploration technology, although we can observe that, if the past is an adequate guide, economic conditions and technology will change rapidly in the future. It might very well be, for example, that subsurface exploration technology in the future could be based on remote sensing so that penetration of the surface is no longer required. We therefore do not think that it is feasible to make meaningful predictions about the probability of advertent or inadvertent intrusion.

Based on these findings, we make two observations about how to deal with human intrusion in the Yucca Mountain standard. First, although there is no scientific basis for judging whether active institutional controls can prevent an unreasonable risk from human intrusion, we think that if the repository is built such controls and other activities can be helpful in reducing the risk of intrusion, at least for some initial period of time after a repository is closed. Therefore, although it cannot be proven, we believe that if a repository is built at Yucca Mountain, a collection of prescriptive requirements, including active institutional controls, record-keeping, and passive barriers and markers, will help to reduce the risk of human intrusion, at least in the near term. The degree of benefit is likely to decrease over time. Further, once other knowledge of the repository is lost, passive markers could attract the curious and actually increase the risk of intrusion. Nonetheless, we conclude that the benefits of passive markers outweigh their disadvantages, at least in the near term.

Second, because it is not technically feasible to assess the probability of human intrusion into a repository over the long term, we do not believe that it is scientifically justified to incorporate alternative scenarios of human intrusion into a fully risk-based compliance assessment that requires knowledge of the character and frequency of various intrusion scenarios. We do however conclude that it is possible to carry out calculations of the consequences for particular types of intrusion events, for example drilling one or more boreholes into and through the repository. We also believe that calculations of this type might be informative in the sense that they can provide useful insight into the degree to which the ability of a repository to protect public health would be degraded by intrusion.

For these reasons, to address the human intrusion issue on an adequate basis, we recommend that the repository developer should be required to provide a reasonable system of active and passive controls to reduce the risk of intrusion in the near term and that EPA should specify in its standard a typical intrusion scenario to be analyzed for its consequences on the performance of the repository. Such an analysis will provide useful quantitative information that can be meaningful in the licensing process, as described later in this chapter. Because the assumed intrusion scenario is arbitrary and the probability of its occurrence cannot be assessed, the result of the analysis should not be integrated into an assessment of repository performance based on risk, but rather should be considered separately. The purpose of this consequence analysis is to evaluate the resilience of the repository to intrusion.

Although we believe that a requirement based on analyses of intrusion consequences is useful in assessing repository performance at Yucca Mountain, such analyses are likely to be more meaningful in selecting among alternative sites (such as by avoiding sites that have potentially valuable mineral, energy, or ground-water resources) than in assessing the performance of a particular site and design. However, Yucca Mountain has already been selected for evaluation as a potential repository site, so the value of analyses of the consequences of human intrusion at Yucca Mountain is limited. Consideration of analytic approaches that would discriminate among alternative sites with greater or lesser likelihood or consequences of intrusion is beyond our charge.

In the remainder of this chapter, we present our argument for the usefulness of an analysis of consequences of a simple intrusion scenario; and provide additional detail on the factors we considered in arriving at our conclusions.

The Consequences of Intrusion

As noted earlier, the consideration of human intrusion cannot be integrated into a fully risk-based standard because the results of any analysis of increased risk as a consequence of intrusion events would be driven mainly by unknowable factors. We reach this conclusion specifically because the numerical value of the risk of adverse health effects due to intrusion is always the product of two factors, the frequency of an intrusion scenario and the measure of consequence. However, the frequency of an intrusion scenario in the distant future is indeterminate.

Technical basis

Some factors affecting an analysis of the consequences of human intrusion can be assessed from a technical base, and some cannot. The historical record of intrusion in the region of the site, including both rate and characteristics (drill depth, hole size, etc.) and the characterization of known mineral and other current resources near the site, can be assessed very well. However, the relevance of the historical record is doubtful. The physical consequences, in terms of the release and probable dispersion of radioactive materials, which is conditional on a defined intrusion scenario — either benevolent or malevolent in purpose — (such as the timing and physical characteristics of the intrusion and whether the intrusion is recognized and remediated), can be assessed moderately well within limits imposed by the level of detail contained in the modeling. Adverse consequences from a specified type of intrusion to a specified local society can also be assessed moderately well, but this assessment for the distant future requires making assumptions about many aspects of the future society, including its sources and technologies for distributing drinking water and food, the ability to detect contamination of food or water, locations of future populations, etc. which cannot be accurately predicted. These assumptions, discussed in Chapters 2 and 3, are inherent in any health-based standard, and we have recommended that for the purposes of compliance analysis they be made explicit through the rulemaking process.

Factors that cannot be technically assessed include the likelihood that institutional controls will persist and succeed over time, or that markers or barriers would persist, be understood, and deter intrusion; the probability that a future intrusion would occur in a given future time period such as in any one year; and the probability that a future intrusion would be detected and remediated, either when it occurs or later. In addition, we cannot predict which resources will be discovered or will become valuable enough to be the objective of an intruder's activity. We cannot predict the characteristics of future technologies for resource exploration and extraction or whether future practice will include sealing of physical intrusions such as boreholes. Continued developments in current non-invasive geophysical techniques, for example, could substantially reduce the frequency of exploratory boreholes.

Consequence-based analysis

Although it would be desirable if the risks associated with the disturbances to a repository by human intrusion could be integrated into a risk assessment of the undisturbed repository performance, technically it is not appropriate to do so. Rather than a complete risk analysis, one alternative is to examine the site- and design-related aspects of repository performance under an assumed intrusion scenario to inform a qualitative judgment. In this approach, the objective would be to perform a consequences-only analysis without attempting to determine an associated probability for the analyzed scenario. We recommend that the Yucca Mountain standard require such an analysis.

We considered at some length the question of whether the calculation of consequences for one or more specified human intrusion scenarios, absent their associated probabilities, could form a useful basis for evaluating a proposed repository site and design. We conclude that the calculations of consequences would provide useful information about how well a repository might perform after an intrusion occurs. The key performance issue is whether the repository would continue to be able to isolate wastes from the biosphere, or if its performance would be substantially degraded as a consequence of an intrusion of the type postulated.

Because the form and frequency of intrusions cannot be predicted, certain assumptions must be made in order to assess the resilience of the repository to intrusion. As in the case of adopting a model of the biosphere and identifying critical groups, selecting an intrusion scenario for analysis entails judgment. To provide for the broadest consideration of what scenario or scenarios might be most appropriate, we recommend that EPA make this determination in its rulemaking to adopt a standard. In this regard, we suggest the following starting point.

For simplicity, we considered a stylized intrusion scenario consisting of one borehole of a specified diameter drilled from the surface through a canister of waste to the underlying aquifer. One can always conceive of worse cases, such as multiple boreholes with each penetrating a canister, but this single-borehole scenario seems to us to hold the promise of providing considerable insight into repository performance with the minimum complication.[1]

An example of a scenario that we believe provides a reasonable basis for evaluation would postulate current drilling technology but assume sloppy practice, such as not plugging the hole carefully when abandoning it, after which natural processes would gradually modify the hole. Although the time at which the intrusion occurs in the future is arbitrary in any hypothetical scenario, we believe it is useful to assume that the intrusion occurs during a period when some of the canisters will have failed but the released materials would not otherwise have had time to reach the ground water. This assumption places emphasis in the consequence analysis on the creation of enhanced pathways to the environment (both to the atmosphere and to the aquifer) as opposed to emphasis on the intrusion's breaching of the canister, which will happen eventually even without human intrusion.

Having defined the reference scenario, the principal questions are what consequence should be assessed and how the result should be interpreted. In our view, the performance of the repository, having been intruded upon, should be assessed using the same analytical methods and assumptions, including those about the biosphere and critical groups, used in the assessment of the performance for the undisturbed case. This analysis should be carried out to determine how the hypothesized intrusion event affects the risk to the appropriate critical groups. We propose that the figure-of-merit for this calculation should be the same as in the undisturbed case, because a repository that is suitable for safe, long-term disposal should be able to continue to provide acceptable waste isolation after some type of intrusion.

The result of this calculation, however, would be a conditional risk: that is, a risk assuming that the hypothesized intrusion occurs. Because the probability is inherently unknowable, we are led to the conclusion that the most useful purpose of this type of analysis is to identify the incremental effects from the assumed scenario. As indicated earlier, we believe that

---

[1] Under many conditions, the effect of multiple boreholes presumably would be the sum of the effects of each taken separately, but circumstances when this assumption is invalid can easily be conceived. Because construction of scenarios is arbitrary, we would argue for the simplest case that tests the repository.

since human intrusion of some type might be likely at some time in the future, a repository should be resilient to at least modest inadvertent intrusions. Because whether and how frequently intrusion events might occur are unknowable, how important these effects are for our expectation that the repository will protect the public can also only be a matter of judgment. Our recommendation is that EPA should require that the conditional risk as a result of the assumed intrusion scenario should be no greater than the risk levels that would be acceptable for the undisturbed-repository case. The conditional risk calculation would not include risks to the intruder or those arising from the material brought directly to the surface as a consequence of the intrusion. As with other policy-related aspects of our recommendations, we note that EPA might decide that some other risk level is appropriate.

Finally, we wish to reiterate that the single borehole scenario that we have discussed should not be interpreted as an estimate of the likely form or frequency of intrusion. A calculation of consequences for such an intrusion removes from consideration a number of imponderables, each of which would otherwise need to be treated separately, including the probability that an intrusion borehole would intersect a waste canister, the probabilities of detection and remediation, and the effectiveness of institutional controls and markers to prevent intrusion. This scenario should not be interpreted as either an optimistic or pessimistic estimate of what might actually occur, because there might be no boreholes that intercept canisters, or there might be more than one. We believe that the simplest scenario that provides a measure of the ability of the repository to isolate waste and thereby protect the public health is the most appropriate scenario to use for this purpose.

ADDITIONAL BASES FOR OUR RECOMMENDATION

In this section we discuss two additional aspects of the human intrusion question that underlie our thinking: the various categories of future human intrusion scenarios and the categories of hazards that could result from a typical borehole intrusion.

Categories of Future Human Intrusion Events

For the purposes of considering how to deal with human intrusion in the context of standard-setting and licensing, we have focused on the particular class of cases in which the intrusion is inadvertent and the intruder does not recognize that a hazardous situation has been created.

We considered several other categories of intrusive events. One case is when the intrusion is inadvertent, but the intruder recognizes that a radioactive waste repository has been disrupted and takes corrective actions. On the assumption that the corrective measures taken are effective and the repository is sealed, this class is not of concern. If, however, corrective actions are not taken or are ineffective, this type of intrusion is operationally the same as the inadvertent intrusion that is not recognized as hazardous, which is the class of cases on which we have focused.

We also considered intentional intrusion for either beneficial or malicious purposes, but concluded that it makes no sense — indeed it is presumptuous — to try to protect against the risks arising from the conscious activities of future human societies. Given the potential energy value of the wastes intended for Yucca Mountain, however, this category of intrusion scenarios might be likely.

Categories of Hazards Resulting From an Intrusion

We have identified three broad types of hazards from radioactive material that could occur as a result of an intrusion into the repository of the type characterized by borehole scenarios. The categories are:

- Hazards to the intruders themselves (the drillers, miners, or handlers of material previously in the undisturbed repository).
- Hazards to the public from any material brought directly to the surface by the intrusive activity. These hazards would arise because such material, now no longer at depth within the repository, would now be mobile in the biosphere, and the public (in addition to the intruders) can be exposed to the material.

- Hazards that arise because the integrity of the repository's engineered or geologic barriers have been compromised by the intrusion.

In the first and second instances, we concluded that analyzing the risks to the intrusion crew and the risks from any material brought directly to the surface as a consequence of intrusion is unlikely to provide useful information about a specific repository site or design and therefore should not provide a basis for judging the resilience of the proposed repository to intrusion. Whenever highly dangerous materials are gathered into one location and an intruder inadvertently breaks in, that intruder runs an inevitable risk. This is not unique to a deep geologic repository, and all deep geologic repositories have this feature. In particular, for inadvertent human intrusion, we believe that it would not be feasible to take regulatory actions today to protect the intrusion crew itself against the risks of its actions, except that requirements identified above associated with active or passive institutional controls might be helpful in this regard.

However, it is possible that an inadvertent intruder would not recognize or would irresponsibly ignore the hazard and would leave the cuttings on the surface so that further exposures would occur. This is the second category of hazards listed above. Our view is that the amount of such future cuttings might not be very different from one repository site or design to another, especially given the unknown nature of an intrusion. Analysis of this hazard too, therefore does not provide information that is useful for judging the ability of the particular repository site and design to protect the public. In this case, we also believe that it is not feasible to take regulatory actions today to alter the repository design to minimize these risks.

We therefore, recommend that the compliance analysis should concentrate on the third category of hazard posed by human intrusion, the one resulting from modification of the repository's barriers and the consequences of these modifications for the ability of the repository to perform its intended function.

5

IMPLICATIONS OF OUR CONCLUSIONS

Early in this study, we were asked by EPA to provide a description of how the form of the standard that we recommend differs from that of the current EPA standard for high-level radioactive waste in 40 CFR 191 and, where there were significant differences, to provide an explanation of the basis for the differences. We have tried to do so in the detailed discussions of Chapters 2, 3, and 4. The purpose of this chapter is to provide a comparison of our recommended approach with 40 CFR 191, including both common elements and differences. It is our intention that this chapter provide a concise summary of what we propose should be done differently and what elements of the 40 CFR 191 approach we recommend be retained.

In addition, we discuss the approach recommended here and that of technology-based standards such as the USNRC's 10 CFR 60. Because our approach is risk-based, it is not useful to make a direct comparison with 10 CFR 60. We do discuss here some aspects of technology-based standards, including ALARA and technology requirements to minimize early releases. Finally, we note some possible administrative consequences of our recommendations.

COMPARISON WITH 40 CFR 191

40 CFR 191 applies to the Waste Isolation Pilot Plant (WIPP) not to the proposed Yucca Mountain repository. Whether some other future repository would be subject to 40 CFR 191 depends on the legislative means taken to initiate it. The 40 CFR 191 standard has three major elements: containment requirements, individual dose limits, and groundwater protection requirements. Section 801 of the Energy Policy Act of 1992 directs EPA to issue a standard to protect the public from radionuclide releases at Yucca Mountain, and requires that the standard be stated in terms of the maximum annual dose equivalent to individual members of the public.

Considerations

We believe that there are two major considerations that give rise to differences between our recommendations and 40 CFR 191.

Generic vs. site-specific standards

By law, EPA is charged with issuing generally applicable standards for protection of health and the environment, and for that reason, 40 CFR 191 is a generic standard. This means that 40 CFR 191 contains provisions applicable for all conceivable terrestrial deep geologic repository sites and types. In addition, at the time that 40 CFR 191 was drafted, the major effort towards establishing a repository was site selection, and 40 CFR 191 was developed to give guidance regarding the feasibility of different types of sites. In contrast, our recommendations concern a standard for the proposed repository at Yucca Mountain. Consequently, we have not addressed site selection, nor have we emphasized potential elements of a standard that would be operationally insignificant at Yucca Mountain. For example, our finding that a containment requirement or release limit is inappropriate is a finding specific to a Yucca Mountain repository; for another geologic setting, we might or might not have reached a different conclusion. The distinction between a generic standard and a site-specific one should be noted as our recommendations are compared with 40 CFR 191.

Dose vs. risk

40 CFR 191 limits individual doses from the undisturbed performance of a repository to 0.15 mSv per year (15 mrem per year). In contrast, we have recommended an approach based on individual-risk limits. Among the reasons why we have chosen risk as the regulatory basis rather than dose, two are important for this discussion. The first is that changes in our understanding of radiation health risks can be accommodated without revision of the level of the standard. If, in the future, scientific evidence becomes available indicating that radiation is

more or less hazardous than our current scientific understanding suggests, the framework we propose would incorporate that new information without a required revision to the level of the standard. The second reason that we have recommended a risk basis is that the probabilities associated with various elements of the exposure calculation can be considered. Our recommended approach is a risk limit based on the probabilistic distribution of a dose and the probability of health effects associated with that dose.

Because the individual dose requirements of 40 CFR 191 have not been implemented, it is not possible to tell whether or how probabilities would be incorporated into estimation of dose. Because the effort at EPA with 40 CFR 191 implementation is now focused on WIPP, and because the individual dose limit is not a particularly important component of the standard for WIPP, it is not clear to us how EPA will interpret its dose limit. In any event, our proposal is clear with respect to our intention that the standard should include consideration of the probabilistic aspect of future exposures.

Differences From 40 CFR 191

What follows is a brief summary of the differences between our recommendations and 40 CFR 191.

Time period

Perhaps the most significant difference between our recommendations and 40 CFR 191 concerns the time period over which the standard is applicable. In 40 CFR 191, the standard applies for a period of 10,000 years. In our proposal, we have specified that the basis for the standard should be the peak risk, whenever it occurs[1]. Based on performance assessment calculations provided to us, it appears that for some reasonable combinations of parameters, peak risks are likely to occur after 10,000 years.

---

[1] Within the limits imposed by the long-term stability of the geologic environment.

Population health effects and release limits

A major element of 40 CFR 191 is its containment requirement, which limits releases of radionuclides to the accessible environment during the first 10,000 years of operation. The stated goal of the release limit was to limit cancer deaths to the general population to 1,000 over 10,000 years. This requirement was to be implemented through a comparison of calculated releases of radionuclides with a table of allowable release limits for each radionuclide. For reasons stated in Chapter 2, we do not think that such a requirement would provide additional protection over that provided by the individual-risk limit for a repository at Yucca Mountain, and we do not recommend that a release limit be adopted.

A related topic is our recommendation in Chapter 2 to employ the concept of a negligible incremental risk, which is the level of risk that can, for radiation protection purposes, be dismissed from consideration. Persons in some local populations outside of the critical group at Yucca Mountain might be exposed to risk from repository releases in excess of the level of negligible incremental risk. However, as individuals, these persons would be exposed to less risk than the risk limit established by the standard for the critical group. On a collective basis, the risks to future local populations are unknowable. We conclude that there is no technical basis for establishing a collective population-risk standard that would limit risk to the nearby population of the proposed Yucca Mountain repository.

Radiation releases from a Yucca Mountain repository can, in principle, be distributed beyond a local population to a global population. In general, the risks of radiation produced by such wide dispersion are likely to be several orders of magnitude below those to a local critical group.

Human intrusion

Under 40 CFR 191, an assessment must be made of the frequency and consequences of human intrusion for purposes of demonstrating compliance with the containment requirements. Human intrusion is not a consideration for compliance with the individual dose limits of ground-water protection requirements. In recognition of the substantial uncertainties involved, EPA has provided detailed guidance for analysis of human intrusion risks and is proposing a reference biosphere be used for the implementation of 40 CFR 191 at WIPP that incorporates an assumption that the future biosphere is much like the present. The EPA requirement includes releases due to drilling cuttings brought to the surface and also includes increases in other radionuclide releases that might occur, for example, through accelerated releases to ground water.

In contrast, we conclude that it is not possible to assess the probability of human intrusion into a repository over the long term, and we do not believe that it is scientifically justified to incorporate alternative scenarios of human intrusion into a risk-based compliance assessment. We do, however, conclude that it is possible to carry out calculations of the consequences for particular types of intrusion events. The key performance issue is whether repository performance would be substantially degraded as a consequence of an inadvertent intrusion for which the intruder does not recognize that a hazardous situation has been created. This consequence assessment is to be done separately from the calculation of compliance with the risk limit from other events and processes, and is to exclude exposures to drillers or to members of the public due to cuttings. We recommend that EPA should require that the conditional risk as a result of the assumed intrusion scenario be no greater than the risk limits adopted for the undisturbed-repository case.

Ground-water protection

40 CFR 191 includes a provision to protect ground water from contamination with radioactive materials that is separate from the 40 CFR 191 individual-dose limits. These provisions have been added to 40 CFR 191 to bring it into conformity with the Safe Drinking Water Act, and have the goal of protecting ground water as a resource. We make no such recommendation, and have based our recommendations on those requirements necessary to limit risks to individuals.

Common Elements With 40 CFR 191

Although our recommendations differ from 40 CFR 191, there are also important similarities in approach.

Dose apportionment

In the recently revised 40 CFR 191, EPA has endorsed the dose limit and dose-apportionment recommendations of the ICRP. We support this approach.

Reference biosphere

In view of the almost unlimited possible future states of society and of the significance of these states to future risk and dose, both EPA and we have recommended that a particular set of assumptions be used about the biosphere (including, for example, how and from where people get their food and water) for compliance calculations. Both EPA and we recommend the use of assumptions that reflect current technologies and living patterns.

Exclusion zone

The original standard, 40 CFR 191, contained a provision for an exclusion zone in the immediate vicinity of the repository. The purpose was to provide a boundary for calculating releases. The zone was presumably to be protected from human activity.

In light of our conclusion in Chapter 4 that it is not reasonable to assume that institutional controls can be maintained for more than a few centuries, we also conclude that there is no scientific basis for assuming that human activity can be prevented from occurring in an exclusion zone or that defining such a zone will provide protection to future generations from exposures in the vicinity of the repository. If, as we recommend, human intrusion is treated separately from the performance of an undisturbed repository, it is reasonable in our view to define a region in which human activities are to be regarded as intrusion and to exclude that region from calculation of the undisturbed repository performance. Beyond the repository footprint, however, there seems to be no practical purpose for defining a larger exclusion zone for the form of the standard we recommend. Without either a release limit or a time limit for the standard for undisturbed performance, an arbitrary boundary serves no purpose.

Use of mean values

We recommend that the mean values of calculations be the basis for comparison with our recommended standards.

LIMITS OF THE SCIENTIFIC BASIS

Our assignment has been to assess the scientific bases for a standard to protect the public health from radiation exposures that might result from radionuclide releases associated with a high-level waste repository at Yucca Mountain. In so doing, we have concluded that for some decisions there presently exists a limited scientific basis required to set and administer such a standard. We have explicitly noted these issues in the preceding chapters, and have indicated that they must be decided on a policy, rather than a scientific, basis. This interplay of scientific and policy issues in the standard has two major implications.

First, where we have identified policy issues, we have recommended that sound public policy would have these issues addressed in rulemaking by the appropriate federal agency, EPA or USNRC. The process of addressing these issues by rulemaking or an equivalent procedure must provide a full opportunity for public participation, especially by the citizens of the affected jurisdictions, and allow the agency the flexibility to take a broad range of public opinion into account in its final public policy judgments. We regard these characteristics as essential for the policy judgments that are required in formulating the standard. In contrast, the licensing process is not suited to this policy-making role, but rather is the arena in which compliance with the standard can be tested.

Several times we have identified possible positions that could be used by the responsible agency in formulating a proposed rule, which is often the initial step in the process. Other starting positions are possible, and of course the final rule might differ markedly from the one proposed. We have tried only to illustrate by reference to other authorities or by example that there seems to be a reasonable policy position from which to begin.

The second implication of the limitations that we have identified is that since they represent current gaps in scientific knowledge, it might be possible that some of these gaps and uncertainties might be reduced by additional research. It seems reasonable, therefore, to ask what gaps could be closed by taking time to obtain more scientific and technical knowledge on such matters as the nature of the waste, its potential use, the health effects of radionuclides, the value of waste products for later generations, and the security of retrievable storage containers. New information in these and other areas could improve the basis for setting the standards if, for example, this information reduced the uncertainty about the effects of very low doses of radiation.

Whether the benefit of new information would be worth the additional time and resources required to obtain it is a matter of judgment. This judgment would be strengthened by a careful appraisal of the probable costs and risks of continuing the present temporary waste disposal practices and use of storage facilities as compared with those attaching to the proposed repository. No such comprehensive appraisal is now available. Conducting such an appraisal, however, should not be seen as a reason to slow down ongoing research and development programs, including geologic site characterization or the process of establishing a standard to protect public health.

TECHNOLOGY-BASED STANDARDS

Technology-based standards play an important role in regulations designed to protect the public health from the risks associated with nuclear facilities. The purpose of these standards is typically to help ensure protection by employing the best available technology, considering cost and other factors. Three issues involving technological approaches have been raised in our study, and we comment on them below.

The ALARA Principle

The "as low as reasonably achievable" (ALARA) principle has been a basic feature of radiation protection for nearly 30 years. It is intended to be applied after threshold regulatory limits have been met, and calls for additional measures to be taken to achieve further reduction in the calculated health effects resulting from radiation exposure of workers or of a population so that final exposures are "as low as reasonably achievable taking account of economic and social factors." ALARA requires a balancing of costs and benefits.

While ALARA continues to be widely recommended as a philosophically desirable goal, its applicability to geologic disposal of high-level wastes is limited at best because the technological alternatives available for designing a geologic repository are quite limited (IAEA, 1989). Further, the difficulties of demonstrating technical or legal compliance with any such requirement for the post-closure phase could well prove insuperable even if it were restricted to engineering and design issues. We conclude that there is no scientific basis for incorporating the ALARA principle into the EPA standard or USNRC regulations for the repository. However, it is nothing other than sound engineering practice to consider whether reductions in radiation dose or risk can be achieved through engineering measures that can be implemented in a cost-effective manner.

10 CFR 60

If EPA issues a standard based on individual risk, USNRC is required to revise its current regulations embodied in 10 CFR 60 to be consistent with such a standard. One purpose of the existing USNRC regulations is to help ensure multiple barriers within the repository system. The concept of multiple barriers, implemented through subsystem requirements, has its origin in the Nuclear Waste Policy Act of 1982. Recognizing this origin, we nonetheless conclude that because it is the performance of the total system in light of the risk-based standard that is crucial, imposing subsystem performance requirements might result in a suboptimal repository design. Care should be taken to ensure that any subsystem requirements for Yucca Mountain do not foreclose design options that ensure the best long-term repository performance.

For example, in 10 CFR 60, there is a subsystem requirement that "the geologic repository shall be located so that the preemplacement ground water travel time along the fastest path of likely radionuclide travel from the disturbed zone to the accessible environment shall be at least 1,000 years. . ." This regulation was written with the presumption that the repository would be located in a saturated zone. At Yucca Mountain, the repository is being considered for location in the unsaturated zone where there is a direct pathway to the atmosphere. This subsystem requirement has focused attention on the ground water and away from the gaseous pathway.

As an explicit example of suboptimization, it could be that in a specific geologic setting the requirement to keep ground water travel times to the accessible environment above 1,000 years, as required by 10 CFR 60, might have next to no effect on future individual risks. However, such a requirement could force the repository design team to alter the specific location of the emplaced waste to a location that, although it could meet the travel-time requirement, would be less optimal. That is, it could imply greater future individual risks — due to other factors such as, for example, a less optimal gaseous pathway or a different geochemical setting that would lead to higher radionuclide solubilities or lower retardation.

Minimum Early Release

Several persons suggested to our committee the use of a technology-based standard that would specify a strict release limit from an engineered barrier system during the early life of the repository. A representative proposal of this type would permit the release of less than 1 part in 100,000 per year of the radionuclides present at 1,000 years after repository closure. It was suggested that this proposal would be consistent with the essentially complete containment concept of 10 CFR 60, and would result in essentially no public health impact for an initial period of time of 300 to 1,000 years, during which the integrity of the engineered barrier system could be assured with a high level of confidence.

We find that such a limitation on early releases from the repository would have no effect on the results of compliance analysis over the long term. Nevertheless, some members of the committee believe that such a limitation might provide added assurance of safety in the near term. Whether to provide such assurance by using a limitation on early releases is a policy decision that EPA might wish to consider.

ADMINISTRATIVE CONSEQUENCES FOR EPA, USNRC, AND DOE

Our recommendations, if adopted, will imply the development of regulatory and analytical approaches for Yucca Mountain that are different from those employed in the past and from some approaches currently used elsewhere by EPA. We further note that several parameters important in risk-based assessment require determination by rulemaking. Both the change in approach and the time required to develop a thorough and consistent regulatory proposal and to provide for full public participation in the rulemaking process, particularly in devising the biosphere models, identifying the critical groups, and defining intrusion scenarios, will require considerable effort by EPA.

Indeed, this process probably will take more than the year, that is currently provided for in the statute, for EPA to complete development of a Yucca Mountain standard in a technically competent way. Although it is important to obtain a timely result, we also believe it is important that EPA take sufficient time to produce a thorough, competent, and consistent standard. A similar duty is imposed on USNRC to assure that its regulation implementing the EPA standard is not compromised by time constraints.

Although a new standard and its implementing regulations might not be available within the two years envisioned in the Energy Policy Act of 1992, that does not mean that DOE's Yucca Mountain Site Characterization Project cannot proceed usefully in the interim. The site-characterization and iterative-performance assessment efforts can continue in the absence of a promulgated standard. DOE has, in fact, been making progress consistent with our recommendations with its series of total-system performance assessments (TSPAs) and we hope that progress will continue. For example, the TSPA-1993 reports from the Sandia National Laboratory (Wilson et al., 1994) and Intera, Inc. (Andrews et al., 1994) examined the performance measure of radiation dose to a maximally exposed individual, in addition to consideration of normalized cumulative releases as defined by EPA in 40 CFR 191.13. The TSPA has also reported on repository performance for a period of one million years as well as for the 10,000-year period. Both the dose calculation and extension of the time period move in the direction of our recommendations. On the other hand, progress for some aspects of DOE's program might depend on the nature of EPA's promulgated standard. For example, the potential risks to a critical group living near Yucca Mountain cannot readily be assessed until the rules for identifying the critical group are defined.

APPENDIX A

BIOGRAPHICAL INFORMATION ON COMMITTEE MEMBERS

Robert W. Fri, *Chair*, is President of Resources for the Future, an independent nonprofit research organization in Washington, DC, that conducts research and policy analysis on issues affecting natural resources and environmental quality. He received a B.A. in physics from Rice University and an M.B.A. from Harvard University. He has served in government as Deputy Administrator of the U.S. Environmental Protection Agency (1971-73) and Administrator of the Energy Research and Development Administration (1975-77) and been a member of numerous committees advising government and resources industries.

John F. Ahearne is Executive Director of Sigma Xi, The Scientific Research Society. He received his B.S and M.S. degrees from Cornell University and his Ph.D. in plasma physics from Princeton University. His professional interests are risk assessment and science policy. He was a commissioner of the U.S. Nuclear Regulatory Commission (1978-83) and its chairman (1979-81). He is a member of the National Research Council's Board on Radioactive Waste Management and has served on a number of the Council's committees examining issues in risk assessment and the future of nuclear power.

Jean M. Bahr is Associate Professor, Department of Geology and Geophysics, Institute for Environmental Studies, and Geological Engineering Program, at the University of Wisconsin, Madison. She received her B.A. degree in geology and geophysics from Yale University and M.S. and Ph.D. degrees in applied earth sciences (hydrogeology) from Stanford University. She is a member of the National Research Council's Board on Radioactive Waste Management.

R. Darryl Banks is Director of the Program on Technology and the Environment at World Resources Institute in Washington, DC. He received his B.A. degree from Coe College and, as Rhodes Scholar, his

Ph.D. from Oxford University. He has worked in the U.S. Congress as a Congressional Science Fellow (1976-77) and a staff member of the Office of Technology Assessment (1977-78). He worked in the Office of Research and Development of the U.S. Environmental Protection Agency (1978-81) before becoming Deputy Commissioner of the New York State Department of Environmental Conservation (1983-92) where he specialized in hazardous waste management issues.

Robert J. Budnitz has been President of Future Resources Associates, Inc. in Berkeley, California since 1981 before which, he was at the U.S. Nuclear Regultory Commission (1978-1980) and was a member of the technical staff and held several management positions at the Lawrence Berkeley Laboratory of the University of California (1967-78). He received his B.A. degree from Yale University and his Ph.D. in physics from Harvard University. His professional interests are in environmental impacts, hazards, and safety analysis, particularly of the nuclear fuel cycle. He has served on numerous investigative and advisory panels of scientific societies, government agencies, and the National Research Council.

Sol Burstein, is a registered professional engineer and member of the National Academy of Engineering. He retired in 1987 as Vice Chairman and Director of Wisconsin Energy Corporation, the holding company for Wisconsin Natural Gas Company and Wisconsin Electric Power Company, of which he also served as Vice President and Director. His career with Wisconsin Electric spanned 21 years, prior to which he spent over 19 years in engineering design and construction work at Stone & Webster. He currently is an independent consultant. He specializes in utility management and nuclear and mechanical engineering. He received a B.S.M.E. degree from Northeastern University and a D.Sc (hon) from the University of Wisconsin at Milwaukee. He has served on numerous industry and government advisory committees and is a member of the National Research Council's Board on Radioactive Waste Management.

Melvin W. Carter is Neely Professor Emeritus of Nuclear Engineering and Health Physics at the Georgia Institute of Technology. He specializes in public health engineering and radiation protection. He received his B.S. degree in civil engineering and an M.S. in public health engineering from Georgia Institute of Technology and his Ph.D. in radiological health from the University of Florida. Before joining the faculty at Georgia Institute of Technology, he had extensive experience in radiologic health as director of government laboratories, including the National Environmental Research Center in Las Vegas (1968-72). He is a Past President of the International Radiation Protection Association and has served on numerous advisory committee of scientific societies; he is also a member of the National Research Council's Board on Radioactive Waste Management.

Charles Fairhurst is Professor of Civil Engineering at the University of Minnesota in Minneapolis, where he has taught since 1956 after having received his B.Eng and Ph.D degrees in mining from Sheffield University, England. His specialties are rock mechanics and mining engineering. He consults internationally on geologic isolation of radioactive wastes and rock mechanics. He is a member of the National Academy of Engineering and the Royal Swedish Academy of Engineering Sciences. He is also Chairman of the National Research Council's Waste Isolation Pilot Plant Committee.

Charles McCombie is Technical Director of NAGRA, the Swiss Cooperative for the Disposal of Radioactive Waste. He has 25 years experience in the nuclear field, more than 15 of which are in radioactive waste management. He serves on a number of international committees advising European and international organizations on radioactive waste management issues. His formal training is in physics with a B.Sc from Aberdeen University, Scotland, and a Ph.D. from Bristol University, England.

Fred M. Phillips is Professor of Hydrology, Department of Earth and Environmental Science, New Mexico Institute of Mining and Technology. He specializes in isotope hydrology and paleoclimatology. He received his B.A. degree from the University of California at Santa Cruz and his Ph.D. in hydrology from the University of Arizona.

Thomas H. Pigford has been Professor of Nuclear Engineering at the University of California, Berkeley since 1959. He is an international consultant in the geologic disposal of radioactive waste. He specializes in the nuclear fuel cycle, nuclear safety, environmental analysis of nuclear systems, and prediction of the release of radionuclides from buried solid waste and their transport through geologic media. He has received many awards for his achievements in engineering, including the Robert E. Wilson Award and the Service to Society Award from the American Institute of Chemical Engineers, the Arthur H. Compton Award from the American Nuclear Society, and the John Wesley Powell Award from the U.S. Geological Survey. He is a member of the National Academy of Engineering and has served on many of the panels and boards of the National Research Council. He was a member of the Presidential Commission on the Accident at Three Mile Island. He is Scientific Master for the U.S. District Court, Hanford Nuclear Reservation Litigation. He earned a B.S. from the Georgia Institute of Technology and a M.S. and Sc.D. in chemical engineering from the Massachusetts Institute of Technology.

Arthur C. Upton is Professor Emeritus of Environmental Medicine at the New York University School of Medicine and currently is Clinical Professor of Environmental and Community Medicine at the Robert Wood Johnson Medical School as well as Clinical Professor of both Pathology and Radiology at the University of New Mexico School of Medicine. He is a member of the Institute of Medicine and has served on numerous committees of the National Research Council, prominently including the series on biological effects of ionizing radiation. He received a B.A. and M.D. from the University of Michigan.

Chris G. Whipple is Vice President of ICF Kaiser Engineers in Oakland, California. He holds a B.S. degree from Purdue University and a Ph.D degree in engineering science from the California Institute of Technology. His professional interests are in risk assessment, and he has consulted widely in this field for private clients and government agencies. Prior to joining ICF Kaiser Engineers, he conducted work in this and related fields at the Electric Power Research Institute (1974-90). He served on the National Research Council's Board on Radioactive Waste Management from 1985 to 1995, and as its Chair from 1992 to March 1995.

Gilbert F. White is Emeritus Distinguished Professor of Geography and Emeritus Director of the Institute of Behavioral Science at the University of Colorado in Boulder. He is a specialist in the social and economic aspects of natural hazards, particularly those associated with water resources. He has served in many government and academic posts, including as President of Haverford College (1946-55) and Professor of Geography at the University of Chicago (1956-69) before joining the University of Colorado in 1970. He is the recipient of many awards, including the Tyler Prize for Environmental Achievement (1987) and the Hubbard Medal of the National Geographic Society (1994). He has served on numerous advisory committees for scientific societies, governments, and the National Research Council. He also chaired the Technical Review Committee on Socio-Economic Effects of Nuclear Waste Disposal for the State of Nevada. He received S.B., S.M., and Ph.D. degrees in geography from the University of Chicago and is a member of the National Academy of Sciences and a foreign member of the Russian Academy of Sciences.

Susan D. Wiltshire is Vice President of JK Research Associates, Inc. in Beverly, MA. She specializes in public policy formulation, strategic planning, and citizen and community involvement in technical programs. She has been a member of a number of National Research Council committees, including the Board on Radioactive Waste Management, and has been president of the League of Women Voters of Massachusetts. She serves on advisory committees to the U.S. Environmental Protection Agency and the National Council on Radiation Protection and Measurements. She holds a B.Sc. degree in mathematics from the University of Florida.

APPENDIX B

CONGRESSIONAL MANDATE FOR THIS REPORT

Letter of J. Bennett Johnston to Robert W. Fri, May 20, 1993

Energy Policy Act of 1992 (P.L. 102-486)
Section 801

Excerpts from the Conference Report (Cong. Rec. H-12056)

United States Senate

COMMITTEE ON
ENERGY AND NATURAL RESOURCES

WASHINGTON, DC 20510-6150

May 20, 1993

Robert W. Fri
Chairman
Committee on Technical Bases
for Yucca Mountain Standards
National Research Council
2101 Constitution Avenue, N.W.
Washington, D.C. 20418

Dear Dr. Fri:

Thank you for the invitation to participate in the initial meeting of the National Academy of Sciences' Committee on the Technical Bases for Yucca Mountain Standards to share my views with the committee about its charge. I regret that I am unable to attend, but I would like to offer the following comments to the committee for its consideration.

I am concerned that the past efforts to set standards to protect the public health and safety at Yucca Mountain have strayed beyond what can be justified based on scientific understanding and principles. The release limits for carbon-14 contained in the 1985 standards are the most obvious example of this problem. It is extremely important that the standards developed for nuclear waste disposal be reasonable, justifiable, and understandable. These standards must be developed based up on a scientific evaluation of the risk involved and must be grounded in the best available scientific data.

Your report will be most helpful if it clearly delineates the technical assumptions, principles, and data that underlie alternative approaches to regulation in as straightforward language as possible. Your guidance on how to apply known scientific principles and how to make judgments where there are technical and scientific uncertainties will be extremely important.

I believe that your committee has a vitally important role to play in bringing the best scientists together to consider these issues and in assuring that reasonable and rational advice is provided to the Environmental Protection Agency. In developing the nuclear waste provisions of the Energy Policy Act of 1992 as we did, the Congress felt that the National Academy of

Sciences was the most qualified to provide this advice and guidance.

Thank you again for the invitation to participate in the committee's meeting next week in Las Vegas.

Sincerely,

J. Bennett Johnston
Chairman

Text of the Energy Policy Act of 1992

TITLE VIII--HIGH-LEVEL RADIOACTIVE WASTE

SEC. 801. NUCLEAR WASTE DISPOSAL.

(a) Environmental Protection Agency Standards.--

(1) Promulgation.--Notwithstanding the provisions of section 121(a) of the Nuclear Waste Policy Act of 1982 (42 U.S.C. 10141(a)), section 161 b. of the Atomic Energy Act of 1954 (42 U.S.C. 2201(b)), and any other authority of the Administrator of the Environmental Protection Agency to set generally applicable standards for the Yucca Mountain site, the Administrator shall, based upon and consistent with the findings and recommendations of the National Academy of Sciences, promulgate, by rule, public health and safety standards for protection of the public from releases from radioactive materials stored or disposed of in the repository at the Yucca Mountain site. Such standards shall prescribe the maximum annual effective dose equivalent to individual members of the public from releases to the accessible environment from radioactive materials stored or disposed of in the repository. The standards shall be promulgated not later than 1 year after the Administrator receives the findings and recommendations of the National Academy of Sciences under paragraph (2) and shall be the only such standards applicable to the Yucca Mountain site.

(2) Study by National Academy of Sciences.--Within 90 days after the date of the enactment of this Act, the Administrator shall contract with the National Academy of Sciences to conduct a study to provide, by not later than December 31, 1993, findings and recommendations on reasonable standards for protection of the public health and safety, including--

(A) whether a health-based standard based upon doses to individual members of the public from releases to the accessible environment (as that term is defined in the regulations contained in subpart B of part 191 of title 40, Code of Federal Regulations, as in effect on November 18, 1985) will provide a reasonable standard for protection of the health and safety of the general public;

(B) whether it is reasonable to assume that a system for post-closure oversight of the repository can be developed, based upon active institutional controls, that will prevent an unreasonable risk of breaching the repository's engineered or geologic barriers or increasing the exposure of individual members of the public to radiation beyond allowable limits; and

(C) whether it is possible to make scientifically supportable predictions of the probability that the repository's engineered or geologic barriers will be breached as a result of human intrusion over a period of 10,000 years.

(3) Applicability.--The provisions of this section shall apply to the Yucca Mountain site, rather than any other authority of the Administrator to set generally applicable standards for radiation protection.

(b) Nuclear Regulatory Commission Requirements and Criteria.--

(1) Modifications.--Not later than 1 year after the Administrator promulgates standards under subsection (a), the Nuclear Regulatory Commission shall, by rule, modify its technical requirements and criteria under section 121(b) of the Nuclear Waste Policy Act of 1982 (42 U.S.C. 10141(b)), as necessary, to be consistent with the Administrator's standards promulgated under subsection (a).

(2) Required assumptions.--The Commission's requirements and criteria shall assume, to the extent consistent with the findings and recommendations of the National Academy of Sciences, that, following repository closure, the inclusion of engineered barriers and the Secretary's post-closure oversight of the Yucca Mountain site, in accordance with subsection (c), shall be sufficient to--

(A) prevent any activity at the site that poses an unreasonable risk of breaching the repository's engineered or geologic barriers; and

(B) prevent any increase in the exposure of individual members of the public to radiation beyond allowable limits.

(C) Post-Closure Oversight.--Following repository closure, the Secretary of Energy shall continue to oversee the Yucca Mountain site to prevent any activity at the site that poses an unreasonable risk of--

(1) breaching the repository's engineered or geologic barriers; or

(2) increasing the exposure of individual members of the public to radiation beyond allowable limits.

Text of Conference Report
[CR page H-12056]

TITLE VIII--HIGH-LEVEL RADIOACTIVE WASTE

Section 801 addresses the Environmental Protection Agency's (EPA) generally applicable standards for protection of members of the public from release of radioactive materials into the accessible environment as a result of the disposal of spent nuclear fuel or high-level or transuranic radioactive waste. Administrator's authority to establish these standards is embodied in section 161b. of the Atomic Energy Act of 1954, Reorganization Plan No. 3 of 1970, and section 121(a) of the Nuclear Waste Policy Act of 1982.

Section 801 builds upon this existing authority of the Administrator to set generally applicable standards and directs the Administrator to establish health-based standards for protection of the public from release or radioactive materials that may be stored or disposed of in a repository at the Yucca Mountain site. The provisions of section 801 make clear that the standards established by the authority in this section would be the only such standards for protection of the public from releases of radioactive materials as a result of the disposal of spent nuclear fuel or high-level radioactive waste in a repository at the Yucca Mountain site. Any other generally applicable standards established pursuant to the Administrator's authority under section 161b. of the Atomic Energy Act of 1954, Reorganization Plan No. 3 of 1970, and section 121(a) of the Nuclear Waste Policy Act of 1982 would not apply to the Yucca Mountain site.

The provisions adopted by the Conferees in section 801 require the Administrator to promulgate health-based standards for protection of the public from releases of radioactive materials from a repository at Yucca Mountain, based upon and consistent with the findings and recommendations of the National Academy of Sciences. These standards shall prescribe the maximum annual dose equivalent to individual members of the public from releases to the accessible environment from radioactive materials stored or disposed of in the repository. The provisions of section 801 do not mandate specific standards but rather direct the Administrator to set the standards based upon and consistent with the findings and recommendations of the National Academy of Sciences.

The Administrator is directed to contract with the National Academy of Sciences to conduct a study to provide findings and recommendations on reasonable standards for protection of the public health and safety by not later than December 31, 1993. In carrying out the study, the National Academy of Sciences is asked to address three questions: whether a health-based standard based upon doses to individual members of the public from releases to the accessible environment will provide a reasonable standard for protection of the health and safety of the general public; whether it is reasonable to assume that a system for post-closure oversight of the repository can be developed, based upon active institutional controls, that will prevent an unreasonable risk to breaching the repository barriers or increasing the exposure of individual members of the public to radiation beyond allowable limits; and whether it is possible to make scientifically supportable predictions of the probability that the repository's engineered or geologic barriers will be breached as a result of human intrusion over a period of 10,000 years. In looking at the question of human intrusion, the Conferees believe that it is also appropriate to look at issues related to predications of the probability of natural events.

In carrying out the study, the National Academy of Sciences would not be precluded from addressing additional questions or issues related to the appropriate standards for radiation protection at Yucca Mountain beyond those that are specified. For example, the study could include an estimate of the collective dose of the general population that could result from the adoption of a health-based standard based upon doses to individual members of the public. The purpose of the listing of specific issues is not to limit the issues considered by the National Academy of Sciences but rather to attempt to focus the study on concerns that have been raised by the scientific community.

Under the provisions of section 801, the Administrator is directed to promulgate standards within one year of receipt of the findings and recommendations of the National Academy of Sciences, based upon and consistent with those recommendations. The Conferees do not intend for the National Academy of Sciences, in making its recommendations, to establish specific standards for protection of the public but rather to provide expert scientific guidance on the issues involved in establishing those standards. Under the provisions of section 801, the authority and responsibility to establish the standards, pursuant to a rulemaking, would remain with the Administrator, as is the case under existing law. The provisions of section 801 are not intended to limit the Administrator's discretion in the exercise of his authority related to public health and safety issues.

The provisions to modify its technical requirements and criteria for licensing of a repository to be consistent with the standards promulgated by the Administrator within one year of the promulgation of those standards. In modifying its technical requirements and criteria, the Nuclear Regulatory Commission (NRC) is directed to assume, to the extent consistent with the findings and recommendations of the National Academy of Sciences, that civilization will continue to exist and that post-closure oversight of the repository will continue, and to include in its technical requirements and criteria, engineered barriers to prevent human intrusion. As with the Administrator, the provisions of section 801 are not intended to limit the Commission's discretion in the exercise of its authority related to public health and safety.

The provisions of section 801 address only the standards of theEnvironmental Protection Agency, and comparable regulations of the Nuclear Regulatory Commission, related to protection of the public from releases of radioactive materials stored or disposed of at the Yucca Mountain site pursuant to authority under the Atomic Energy Act, Reorganization Plan No. 3 of 1970, the Nuclear Waste Policy Act of 1982, and this Act. The provisions of section 801 are not intended to affect in any way the application of any other existing laws to activities at the Yucca Mountain site.

APPENDIX C

A PROBABILISTIC CRITICAL GROUP

Although the components of a probabilistic computational approach have considerable precedent in repository performance, we are not aware that they have previously been combined to analyze risks to critical groups. We have therefore outlined in this appendix a fairly explicit example of how this approach might be implemented for the case of exposure through contaminated ground water. The main purposes of this example are to show that the approach is feasible and to illustrate the steps necessary to perform such a calculation. The example uses a Monte Carlo method for modeling exposure consistent with that employed in the hydrologic modeling of radionuclide transport. In presenting this appendix, we do not intend it as a detailed recommendation, but an exploration of at least the more important issues that are likely to arise in an actual compliance calculation. The additional detail in this appendix is warranted because the technique has not been applied to this problem in the past, as far as we are aware.

The following outline of steps is designed to provide an illustrative example of the types of calculations that could be employed in an exposure scenario analysis. The specific process described here is only one of a variety of alternatives that EPA might consider during its rulemaking. It is based on a number of choices and general considerations, some of which are reviewed below prior to a description of the steps themselves.

a. Technical feasibility of the calculations requires specification of one or more exposure scenarios. As described in Chapter 3, a scenario includes parameter values or distributions that provide quantitative descriptions that include where people live, what they eat and drink, and what their sources of water and food are. A given scenario might include the lifestyle and activities of only farmers or a mix of economic lifestyles and activities of farmers, miners, defense workers, and casino operators, for example. It might be based on actual current activities in the area of interest, on current activities in some adjacent area, or potentially on any number of hypothetical future activities. The only technical consideration in the selection of an exposure scenario is whether the specified scenario provides sufficiently well defined parameters or parameter distributions to make calculations feasible. The selection of the exposure scenario, along with its associated parameter values, is fundamentally a policy choice and therefore an appropriate responsibility of rukemakers. Broad participation in this policy decision by the various affected interested parties and acceptance of the scenario as a reasonable basis for performance assessment are likely to be essential to acceptance of any results of the analysis (NRC, 1993).

b. Even for a narrowly specified set of parameters, it is possible that the calculation procedure can be manipulated to obtain results closer to those desired by the analyst. It might not be possible to eliminate all opportunities for this type of manipulation. However, careful consideration of these possibilities during the rulemaking process might help to develop guidelines for calculations to address some of the potential pitfalls. For example, we were particularly concerncd with avoiding strategies that would reward uncertainty in the temporal or spatial distribution of radionuclides in ground water. A procedure in which larger uncertainty in transport parameters leads to a reduction in calculated risk, relative to the risk that would be calculated were transport parameters less uncertain, would provide a strong disincentive to reduce uncertainty through site-characterization activities. A second issue is how to quantify properly the risk in areas of low-population density, because the probability of an individual receiving a dose in these areas is dependent on whether any individual is present in the area at the time when radionuclides are present in the underlying ground water. A critical feature of this model, therefore, is that a method must be incorporated for calculating the probability that people are present over the contaminated plume of ground water.

c. The method illustrated in this appendix employs a fully probabilistic treatment of all aspects of the exposure scenario. This results in a computationally intensive procedure. It might be possible to reduce the computational requirements by treating parts of the calculation deterministically or analytically.

d. The illustrative example focuses on exposures and risks associated with ground-water use. The fact that gaseous releases are not included in this example should not be interpreted as a judgment that such releases can be excluded from performance assessment and compliance evaluation. A separate exposure scenario, with a different critical group, would be required for evaluation of the gaseous exposure pathway. In the end, however, one pathway will result in the maximum risk and define the critical group whose protection would be the primary metric for setting the standard.

Example Steps Required for Implementation of a Monte Carlo Analysis

Step 1: Identify general lifestyle characteristics of the larger population that includes the critical group.

The first step is to identify the type of people who would be likely to receive the highest doses and therefore be at greatest risk. These people make up a group that might be considerably larger than the critical group, but of which the critical group will be a subset. As noted earlier, this step involves subjective choices that should be part of the rulemaking process. For purposes of illustration, this example assumes a farming community scenario, based on present-day conditions in the Amargosa Valley.

Step 2: Quantify important characteristics, distributions of characteristics, and geographic location of the chosen population.

The second step addresses two aspects of the exposure analysis. First, any analysis of exposure will require specific information on the living patterns, activities and other characteristics of potential members of the exposed population that can be used as input to deterministic or probabilistic simulations. Second, if identification of the characteristics of currently occupied land and technologies (such as soil type, slope, depth to ground water, well depth, etc.) provides a technical basis for limiting the simulation area for exposure analysis, significant reduction in the computational effort required for the calculations would result.

In a Monte Carlo simulation, each of the pertinent parameters is represented by a distribution of values, from which one value for each is randomly selected for each of many calculations. For the purpose of this example, we assume that each of these factors could be quantified using surveys and studies of the existing population in the region. Correlations between factors would need to be identified, such as relationships between farm density and soil type or depth to ground water. Analysis of these data would provide a basis for a model of the farming economy that can be used to identify geographic areas in the basin that have the potential for farming and ground-water use. It is important to note that these areas would not necessarily correspond to the current areas of highest population density or water use, since there might be areas of arable land that have not been developed due to restricted access (anywhere in the Nevada Test Site, for example). There might be areas where higher rates of water use could be easily sustained but have not been implemented by some farmers, or for a variety of other reasons.

Step 3: Simulation of radionuclide transport and identification of potential exposure areas

The third step is to identify the potential intersections of potentially farmable areas and areas beneath which radionuclide-contaminated ground water occurs. Delimiting the intersections of these areas can further reduce the computational effort.

The physical location and chemistry of the plume of contamination can be identified by performing a series of Monte Carlo simulations of the release and transport of the wastes through the unsaturated zone to the water table and in the saturated zone. Each simulation will generate a plume path (direction, width, depth below the water table, thickness) and its surface footprint. This footprint can be overlaid on the map of potential farm density or water use to determine a potential exposure area. If the model employs an appropriate sampling of the input parameters controlling radionuclide release and transport, each of the many plume realizations can be considered an equally likely outcome of radioactive waste disposal at Yucca Mountain. If the number of plume simulations is sufficiently large, the series of calculations defines the statistical characteristics of the problem.

Step 4: For each plume realization, identify critical "snapshots" of radionuclide distribution at time(s) for which the plume underlies exposure area(s) identified in step 3.

Even if the plume evolution were perfectly predictable, and hence the potential exposure area perfectly constrained, not all inhabitants of this exposure area would be at risk. There will be a long period of plume history (that does not even begin until radionuclides reach the saturated zone) during which radionuclide contaminated ground water will not have reached the aquifer beneath a potential exposure area. Inhabitants of a potential exposure area living there during these periods are at no risk. Once the plume reaches the aquifer beneath an exposure area, the risk to inhabitants will vary with time as the areal extent of the plume and radionuclide concentrations in the contaminated ground water change during plume migration. If the critical group comprises a set of individuals who have the greatest average risk, then the temporal as well as spatial distribution of risk must be considered in identifying the group. The purpose of this step is to account for the temporal variation in risk by identifying a) the time at which inhabitants of a potential exposure area will be at maximum risk and b) the corresponding radionuclide distribution in ground water at that time. The subsequent exposure analysis can then be conducted employing the radionuclide distribution for this critical time.

Each of the simulations produces a realization of plume evolution in space and time. The spatial distribution of radionuclide concentrations in ground water at an instant in time constitutes a plume snapshot. If rates of plume evolution are slow, as would be expected from performance assessment calculations conducted to date for Yucca Mountain, a snapshot for an instant in time is also likely to be representative of the plume distribution over the course of a human lifetime, or even over many generations. Examining a series of snapshots generated by a simulation, one can identify the period(s) of time, for each simulation, during which peak radionuclide concentrations or high total (volume integrated) activities are present beneath the area(s) delimited in step 3. These periods should correspond to the times at which the population in the exposure area would be at significant risk. Determining the time of greatest risk might not be straightforward, however, because times of peak concentration (possibly over a very limited area) might not coincide with times of greater plume extent, that would have somewhat lower concentrations but greater total activity.

Step 5: Generate exposure realizations

Having identified the time period of maximum potential exposure for each plume realization, it is also necessary to determine the spatial distribution of potential doses and health effects to identify the critical group and to calculate the risk to an average individual in that group. The next step, then, is to use the plume snapshots in the Monte Carlo series of exposure simulations.

For each of the plume snapshots selected in step 4, a large number of Monte Carlo simulations would be performed. For each exposure simulation, statistical distributions of population characteristics as determined in step 2 would be sampled to generate a distribution of farms with associated inhabitants, wells, crops, livestock, and support services within and surrounding the exposure area (as determined in step 3). Well depth and screened interval, rates of water use, food sources and consumption rates, etc. would also be determined by sampling from the parameter distributions. The number of exposure simulations must be large enough to produce an adequate sampling of exposure parameter distributions.

Each simulation should cover a large enough region outside the exposure area to allow adequate definition of dose variations between the exposure area and the surrounding region. Exposures outside the area overlying the plume could result from local export of water or food from the exposure area, factors that must be included in the exposure analysis. Some exposures might also occur to inhabitants living over the plume but outside areas of intense farming or water use.

Step 6: Calculation of dose distributions for exposure realizations

The spatial relations between plume boundaries and well locations in the exposure realizations will determine which wells have the potential, constrained by well depth and screened interval, to produce water leading to human exposures. For a known concentration, rates of water use for drinking and irrigation will determine the activity extracted from the ground, and the subsequent distribution of that activity to humans, crops, livestock, etc., and the resulting dose to each inhabitant represented in the exposure realization.

Step 7: Interpretation of exposure simulation results to identify critical subgroups

For each of the plume realizations, the results of the exposure simulations can be combined to yield a spatial distribution of expected dose, which can then be used to identify the geographic area inhabited by the critical subgroup for a given plume realization.

For example, the individual doses of the combined plume and exposure simulations could be divided into subsets based on geographic location of the inhabitants. The sizes of the subareas should be adjusted to provide adequate resolution of the spatial variation in individual dose and to account for the variations in the scenario-specific population density over the simulation region. This could result in a highly variable grid size. A sufficient number of individuals must be simulated in each subarea to allow computation of a meaningful average dose. For each subarea, an average individual dose could be computed as the arithmetic mean of the individual doses in that subarea generated by the exposure simulations. The product of this average dose and the factor relating doses to health effects (5×10^{-2} fatal cancers/Sv) would be the average lifetime risk for an individual in the subarea.

The procedure for identifying the critical subgroup for one of the plume realizations would begin by delineating the subarea of the simulation region with maximum average risk plus additional subareas in which the risk is greater than or equal to one-tenth the risk in the subarea with maximum risk. These subareas constitute a trial area for a critical subgroup that is homogeneous with respect to risk. The average risk in this trial area is calculated as the arithmetic mean of the subarea risks. A critical sub-group can be considered homogeneous if it satisfies the criteria detailed in Chapter 2.

Step 8: Calculation of average risk to members of the critical group

The procedure outlined in step 7 will generate a risk for the critical subgroup corresponding to each of the plume realizations. The arithmetic average of these critical subgroup risks over all plume realizations is the technically appropriate representation for the critical-group risk. The variability in risks between critical subgroups is related primarily to the variability in potential plume concentrations and locations resulting from the probabiliistic simulations of release and transport mechanisms. Using the average critical subgroup risk provides an estimate of the risk to the critical group exposed to the average plume. Additional insight might be obtained by examining the cumulative distributions of the critical subgroup risks.

APPENDIX D

THE SUBSISTENCE-FARMER CRITICAL GROUP

In Chapter 2 we recommend that the form of the standard be a limit to the risk to the average individual in a future critical group. This appendix summarizes the steps that could be involved in assessing compliance with such a standard for a particular exposure scenario that defines the critical group as including a subsistance farmer exposed to a maximum concentraton of radionuclides in ground water.

The risk involved here is the risk of ill health from a radiation dose. Risk entails probabilities as well as consequences. A risk analysis must entail the development of probabilistic distributions of doses to future individuals for various times in the future and the development of probabilistic distributions of consequences (health effects) from those doses[1].

There are various means of constructing risk measures from such probabilistic distributions to be compared with a risk limit. The risk measure recommended in Chapter 2 is the expected value of the consequences, determined by integrating the probabilistic distribution of consequences over the entire range of estimated consequences.

The conceptual approach to analyzing risks to future individuals from a geologic repository will be illustrated here for undisturbed performance (e.g., not including human intrusion, meteoric impact, etc.). Radionuclides can be released via air or water pathways. The steps in calculating risks for the water pathways are summarized here. Similar steps are involved in calculating risk to future individuals via air pathways. For this illustration, radionuclides in waste solids are calculated to eventually dissolve in water and undergo hydrogeologic transport to the saturated zone and subsequently transport via an aquifer to the biosphere. A plume of contaminated ground water will spread out underground, downstream from Yucca Mountain, to places where it might be susceptible to human use. Calculating the space- and time-dependent probabilistic

---

[1] A probabilistic distribution of a variable can be thought of as the probability per unit increment of that variable as a function of that variable.

distributions of concentrations of radionuclides in the ground-water plume is the purpose of geosphere performance analysis.

Calculation of Geosphere Performance

As described in Chapter 3, there are many different possible mechanisms and pathways for the dissolution-transport processes. For example, dissolved radionuclides might be transported to the lower aquifer by slow processes that provide time for local sorptive equilibrium with the rock. In other locations, radionuclides might be transported via fast pathways resulting from episodic local saturation, with little time for diffusion into the surrounding rock matrix.

The analysis must begin with what might be, in principle, a time-dependent statistical distribution of such scenarios of release and transport. Enough scenarios must be identified that will reasonably sample the events that can contribute to important releases of radionuclides. The probability of each of these geosphere scenarios must be estimated so that the resulting analysis can reasonably approximate the statistical distribution of consequences that would be expected.

For each geosphere scenario there are large uncertainties in the parameters used in the equations for release and transport. For full probabilistic analysis, a state-of-knowledge distribution for each parameter must be developed. Using the equations of transport, these probabilistic distributions of input quantities can be projected into a probabilistic distribution of ground-water concentration, which will vary with position and time. Although many useful calculations are made with analytic techniques (NRC, 1983), detailed results require discretizing input quantities, followed by event-tree transport calculations of a large number of combinations of input quantities (EPRI, 1994) or by Monte Carlo/Latin Hypercube sampling of a smaller number of data combinations, as used by the WIPP and Yucca Mountain Projects (Wilson et al., 1994). Semianalytical adjoint techniques that help create probabilistic distributions from the discretized results are also available. Any of these numerical techniques can yield useful probabilistic distributions, if done properly. The choice is better left to the analyst, who must consider limitations of time, budget, and computer power. Estimates of errors introduced by sampling techniques should be included when such techniques are used to reduce the number of discrete calculations.

These space- and time-dependent probabilistic distributions of concentrations in ground water, with emphasis on ground water beyond the repository footprint, are the input quantities needed for calculating radiation doses, consequences, and risks for the biosphere scenarios. Similar approaches are followed for calculating the space and time dependent concentrations of radionuclides released to the atmosphere.

Many analysts employ system software that feeds geosphere results directly into biosphere calculations, bypassing the display of probabilistic distributions of concentrations in ground water.

Calculation of Biosphere Performance

For the biosphere scenario involving the subsistence-farmer critical group, ground water is assumed to be withdrawn at the location of temporal-maximum concentration of radionuclides. The time of that maximum concentration specifies the time at which the doses, consequences, and risk are being calculated at that location. In the era of temporal-maximum concentration, the concentrations at a given location vary little over a human lifetime, so the ground-water concentration can be assumed constant in calculating lifetime doses and risks for that critical group. The critical assumption in this model, then, is that a subsistence farmer extracts water from the location of maximum concentration of radionuclides in the aquifer, provided that no natural geologic feature precludes drilling for water at that location.

The subsistence farmer is assumed to use the extracted contaminated water to grow his food and for all his potable water. Conservatively, the farmer is to receive no food from other sources. A pumped well to extract ground water can perturb the local flow of ground water, so that concentrations of contaminants in the extracted water can be less than in the unperturbed ground water. The extent of concentration reduction depends on the extraction rate (Charles and Smith, 1991). A reasonable extraction rate can be calculated assuming that the subsistence farmer or even the entire critical group uses a single well for extracting ground water.

If the subsistence farmer's water is obtained from commercial pumping of the underground aquifer at the point of maximum local contamination[2], the effect of commercial rates of water extraction on the withdrawn concentration can be included in the analysis. Obviously, for commercial water withdrawal, it is the withdrawal location rather than the location of the subsistence farmer that is important.

The vertical variation of concentration in ground water at a given surface position can be obtained from the geosphere analysis. If methods of predicting the vertical location of the point of water withdrawal within the aquifer are defensible for the long-term future, then the effect of withdrawing at locations other than that of the vertical maximum concentration can be included. Otherwise, arbitrary assumptions of well depth would diminish confidence in the resulting calculated risk.

The largest radiation exposure to future humans from contaminants in ground water is predicted to result from internal radiation from ingested or inhaled radionuclides. For the water pathways, eating food contaminated by irrigation or by other use of contaminated ground water for growing food is expected to be the source of largest dose, greater than doses from drinking water (NRC, 1983). Therefore, realistic prediction of doses and risks to future humans requires knowledge of their diets and amounts of food and water consumed. Such information for the distant future is unknowable. Therefore, as is done in all other biosphere scenarios, we must assume that future humans have the same diets as ourselves (including food and water consumption). This amounts to the unavoidable policy decision that geologic disposal is to protect future humans whose diets are the same as ours or whose diets would not lead to greater radiation doses from using contaminated water than would the diets of people today.

All biosphere scenarios must also rely on data for the uptake of radionuclides from contaminated water into food. Here, one can rely on scientific data for the typical soil conditions and for the kinds of foods assumed for this analysis. For a given food chain and for drinking, the amount of radioactivity ingested in a given time, or over a human lifetime,

---

[2] There is a current proposal for commercial withdrawal of ground water from the aquifer near Yucca Mountain. This water could be distributed to local communities as well as others that might exist or be developed farther from Yucca Mountain.

is proportional to the concentration of radionuclides in the extracted ground water.[3]

The ingredients of the biosphere approach described here, beginning with specified concentrations in extracted ground water, are identical with those of the widely used GENI computer code developed by Napier et al. (1988). The GENI code is used by the WIPP Project in predicting doses to future individuals who utilize contaminated water for drinking and for growing food and who receive no food from outside sources. It is an example of what could be used or updated for calculating subsistence-farmer doses.

The GENI code includes intake-dose parameters recommended by ICRP and other agencies. Therefore, employing GENI or a similar code to predict radiation doses to future humans who inadvertently use contaminated water requires the additional assumption that future humans have the same dose-response to ingested radioactivity as do present humans. All biosphere scenarios adopt this assumption. Of course, it is expected that the intake-dose parameters will be updated when new information is available.

Given the probabilistic distribution of concentration of radionuclides in extracted ground water at a given future time and location, the human-uptake-response model, such as GENI, can predict the statistical distribution of radiation doses to the subsistence farmer. Because the ground-water concentrations vary little over a human lifetime, it is necessary only to sum the dose commitments for a human who uses that contaminated water over his/her lifetime. The result is a probabilistic distribution of lifetime dose commitments, easily converted to lifetime average annual dose commitments.

The probabilistic distribution of lifetime dose commitments can be converted into a distribution of consequences by multiplying each value of dose commitment by the appropriate dose-risk parameters, obtainable from ICRP and others. If the constant dose-risk parameter of the linear hypothesis is used, the probabilistic distribution of consequences will differ from that of doses by only a constant multiplier. Here, by adopting dose-

---

[3] This assumes uptake factors, i.e., distribution coefficients for a given radiochemical species in a given plant or other organism immersed in contaminated water, that are independent of radionuclide concentration.

risk parameters developed for present humans, we are assuming that future humans will have the same present risk when exposed to a given radiation dose. All biosphere scenarios adopt this assumption. Of course, it is expected that the dose-risk parameters will be updated when new information is available.

Each value of the consequence is then multiplied by the probability distribution function for that consequence, and this integrand is then integrated over all consequences. The result is the calculated risk to the subsistence farmer from ground-water pathways, expressed either as the lifetime risk or as the lifetime average annual risk. To this risk from the ground-water pathways are to be added other calculated risks for the subsistence farmer, who is the individual at maximum risk within the critical group.

To obtain the risk to the average member of the critical group, for compliance determination, it can be arbitrarily assumed for simplicity that there is a uniform distribution of individual risk within that group.[4] Because ICRP's homogeneity criterion specifies that the critical group should have no more than a tenfold variation in individual dose, and because large departures from the linear dose-response theory are not expected for this calculation, the expected value of the risk to the average individual will be about one-half that of the maximally exposed subsistence farmer.[5]

The expected value of risk to the average individual within the subsistence-farmer critical group is to be compared with the risk limit that is to be selected for compliance. The regulator can specify how far below

---

[4] Adopting any distribution, uniform or otherwise, for the risks within a critical group projected to exist in the distant future, ca. 100,000 years and beyond, is **arbitrary**, because the habits, location, etc. of that future group of people are not knowable to us. Whether one postulates some distribution, as is done here, or calculates a distribution based on the assumed relevance of the current site-specific population, adopting any such distribution for the future is arbitrary.

[5] Because of the large uncertainties in the calculated doses and risks to any of these individuals, the uncertainty of uniformity of risk within the group cannot introduce an important uncertainty in the result. An uncertainty of 2 or 3 in the calculated dose is not expected to be important.

or above the specified risk limit the calculated risk must be for compliance decision.[6]

---

[6] UK's NRPB specifies the calculation of a 95% confidence interval for the expected or central value of risk. The upper value of this confidence interval is what is compared with a regulatory limit [Barraclough *et al., 1992]*.

APPENDIX E

PERSONAL SUPPLEMENTARY STATEMENT OF THOMAS H. PIGFORD

INTRODUCTION AND SUMMARY

This supplementary statement clarifies two alternative methods of calculating radiation exposures to people in the far future. They are the exposure scenarios involving the "probabilistic critical group" described in Appendix C and the "subsistence-farmer critical group" described in Appendix D. Both exposure scenarios involve critical groups, as recommended by the International Commission on Radiation Protection (ICRP). ICRP also recommends that the critical group include the person at highest exposure. The objective is to ensure that if the individual at calculated maximum exposure is suitably protected, no other individual doses will be unacceptably high [ICRP, 1985ab].

I believe that this objective can be reasonably met if exposures and risks are calculated using the subsistence-farmer scenario and if the calculated risks meet the Standard's performance criterion. The subsistence-farmer is the individual at calculated maximum risk. Thus, the subsistence-farmer approach is conservative and bounding. Its use represents wide national and international consensus for safety assessment when characteristics of exposed populations are not known. In contrast, the probabilistic critical-group calculation is based on arbitrary choices of reference populations, is not well defined, is not mathematically valid, and is subject to manipulation. It could lead to much lower calculated doses and risks. There is no indication, however, that this country needs to adopt a calculational approach that is so much more permissive than current national and international practice. Its adoption would undermine confidence in the adequacy of public health protection and jeopardize future success of the Yucca Mountain project.

A policy decision common to exposure scenarios in Appendices C and D of the Report is that future humans will have diets and food-water intake similar to that of people now living in the vicinity. In both exposure scenarios, calculations are to be made for future people who do not have extreme sensitivity to radiation, who have the same response to radiation as present people, and who do not have abnormal diets. This Supplementary Statement speaks of calculating maximum and average doses and risks to such future humans, not to persons who may be at greater risk because of unusual diets or unusual sensitivity to radiation.

COMMENTS AND EXPLANATION

1. Among the many possible exposure scenarios, the subsistence-farmer exposure scenario is the most conservative. It is bounding. All future people will be protected if the calculated subsistence-farmer dose/risk meets a prescribed safety limit.

Future humans can be exposed to radiation by drinking well water containing radionuclides and consuming food grown from that contaminated well water.[1] [2] In addition to assuming diets and food-water

---

[1] Calculated concentrations of radionuclides in ground water are a function of location and time. Exposure calculations translate these concentrations into estimates of dose and risk to future people. The method of exposure calculation is the "exposure scenario"; it is sometimes called the "biosphere scenario".

[2] The Committee is also concerned with the persons exposed to "the highest concentration of radiation in the environment". The environment includes air, water, and soil. The radiation in that environment consists of photons, free electrons, and alpha particles from radioactive decay of radionuclides. The "concentrations" of such radiation are rarely calculated, but could be deduced from calculated radiation fluxes. Evidently the Committee has in mind possible exposure from external radiation, such as doses to the skin from swimming in contaminated water or from being immersed in contaminated air. However, studies presented to the Committee show that such doses and risks from external radiation in the environment are minor compared to doses and risks from inhalation and ingestion of radionuclides that may be released to the

(continued...)

intake typical of that of present humans, it is also necessary to assume how much of the lifetime intake of food and water is affected by water contaminated with radioactivity, as well as how near the withdrawal well is to the repository. These "human activity" assumptions are most difficult to deal with.

Future people are deemed to be suitably protected if their calculated lifetime radiation doses and risks are less than a prescribed dose or risk limit. The calculational method should be constructed so that if the person receiving the calculated maximum dose is suitably protected, then all future people with similar diet and dose response will also be protected [ICRP, 1985ab]. To ensure such protection we should assume conservatively that some future individuals are subsistence farmers who use contaminated ground water for drinking and for growing their food over their entire lifetime.[3] To ensure that no future person receives a greater lifetime dose, we assume that the water used by the subsistence farmer is extracted from the location of maximum concentration in ground water.

The subsistence farmer calculation is the most conservative for the type of people assumed for dose/risk calculations. It is bounding. It is patterned from the widespread practice, current and historical, of calculating dose and risk to maximally exposed individuals where the exposure habits of real people cannot be specified or calculated. It is also the most stringent exposure scenario.

(...continued)
environment from a geologic repository [Napier *et al.*, 1988].

[3] Large uncertainties in the calculation of radionuclide concentrations in the geosphere mean that calculated doses and risks to the subsistence farmers will also be extremely uncertain. Consequently, dose/risk estimates will be little affected whether all or only a "substantial portion" of the subsistence farmer's intake of water and food is contaminated by the extracted ground water.

2. There is international consensus to calculate doses and risks for subsistence-farmers in determining compliance with a safety limit for geologic disposal. There is no such consensus for the probabilistic critical group proposed by the Committee.

There is considerable precedence, in the U.S. and abroad, for basing dose and risk predictions on a subsistence farmer, or on a critical group that includes that subsistence farmer, as defined above.[4] Projects for high level waste disposal in the UK, Sweden, Finland, Canada, and Switzerland follow similar practices [Barraclough *et al.*, 1992; Charles et al., 1990; Vieno *et al.*, 1992; Davis *et al.*, 1993]. Switzerland's geologic disposal project defines the critical group as a self-sustaining agricultural community located in the area(s) of the highest potential concentration. Switzerland assumes that no food and water are obtained from outside sources [Switzerland, 1985, 1994; van Dorp, 1994].

In discussing the choice of critical groups and exposure scenarios for long-term waste management, UK's National Radiological Protection Board (NRPB) [Barraclough *et al.*, 1992] states:

> ".... it is appropriate to use hypothetical critical groups. For the purposes of solid waste disposal assessments, these are assumed to exist, at any given time in the future, at the place where the relevant environmental concentrations are highest, and to have habits such that their exposure is representative of the highest exposures which might reasonably be expected."

and, for long-term estimates of radiation dose and risk, Barraclough *et al.*, state:

> " ... the 'reference community' replaces the critical group, and is located so as to be representative of individuals exposed to the greatest risk, at the point of highest relevant environmental concentrations. .... The reference community should normally comprise 'typical' subsistence farmers, i.e., perhaps a few families who produce a range of food to feed themselves."

---

[4] Many of these projects adopt the term "maximally exposed individual" instead of the "subsistence farmer". The dose/risk assumptions are the same.

Likewise, the U.S. Yucca Mountain project estimates radiation doses to future individuals on the basis of conservative subsistence farmers whose entire food and water are contaminated with radionuclides from the proposed repository [Andrews *et al.*, 1994; Wilson *et al.*, 1994]. The GENII code [Napier *et al.*, 1988; Leigh, *et al.*, 1993] is used to define the biosphere scenario and to calculate doses to subsistence farmers.

The U.S. Nuclear Regulatory Commission (USNRC) calculates radiation doses to future individuals who could be affected by geologic disposal [McCartin *et al.,* 1994; Neel, 1995]. To calculate future human exposures, USNRC assumes a hypothetical farm family of three persons who obtain all their drinking water from a contaminated well. Well water is used to grow a large portion of the family's vegetables, fruits and grains. All of the family's beef and milk is obtained from farm animals fed on vegetation irrigated by contaminated well water [Napier, *et al.*, 1988]. The assumed farm family's well is not restricted to the location of the present population[5]. Well depth and withdrawal rate are not constrained by present practice in the vicinity of Yucca Mountain. These assumptions meet the criteria for the conservative subsistence farmer described above. They meet the ICRP criteria for calculating doses for geologic disposal [Neel, 1995].

There are numerous other relevant examples. The U.S. WIPP project to dispose of transuranic waste in bedded salt calculates radiation doses based on a biosphere scenario that is the equivalent of the conservative subsistence-farmer approach. They use the GENII code [Napier, *et al.*, 1988; Leigh, *et al.*, 1993] to calculate individual doses once concentrations in water have been estimated. The estimated doses can be converted to risks by using the dose-risk conversion factors. Sandia National Laboratories recently used the subsistence-farmer calculation to evaluate doses and risks from DOE-owned spent fuel emplaced in a tuff repository [Rechard, 1995]. DOE's Hanford Environmental Dose Reconstruction Project [Farris, 1994ab] adopts variants of the subsistence farmer approach to calculate doses when occupancy factors and locations of actual exposed people are not sufficiently known. When the location, occupancy, and food source of real people cannot be identified, as in specifying a generically safe level in drinking water or in calculating long-

[5] No one in the present population lives nearer than 20 miles from Yucca Mountain.

term performance of geologic disposal, dose/risk estimates are based on the more conservative approach involving the hypothetical maximally exposed individual.

Thus, adopting the subsistence-farmer approach is the consensus among the several geologic disposal projects in other countries and in the U.S., including the USNRC plans for calculating individual doses for a high-level waste repository. It is adopted to calculate doses when actual location and habits of potentially exposed people are not known.

On the other hand, the Committee has identified no reference wherein the kind of probabilistic exposure analysis of future human activities, as proposed in Appendix C, has been adopted for geologic disposal.

3. The reference population for the Committee's probabilistic exposure can be chosen arbitrarily.

The Committee's probabilistic exposure calculations are to be based on extrapolation of location and habits of an arbitrarily selected reference population. The Committee acknowledges (cf. Appendix C) that the selection of the reference population for probabilistic analysis would be arbitrary. The population might be present inhabitants in the vicinity, inhabitants in some adjacent area, or inhabitants of an entirely different community[6], or inhabitants of a hypothetical future population. It could evidently be any population of the past, present, or future. The Committee would only require sufficient parameters to enable a calculation to be made.

The Committee illustrates the probabilistic method by adopting an arbitrary reference population consisting of those people living 20 or more miles away from Yucca Mountain.[7]

---

[6] It has been suggested by proponents of the Appendix C approach that the population of Las Vegas could be a suitable reference population instead of the population in the region surrounding Yucca Mountain.

[7] No people now live nearer than 20 miles from Yucca Mountain because the nearer land is publicly owned.

4. The subsistence-farmer calculation of dose and risk fulfills recommendations of the International Commission on Radiological Protection (ICRP), the probabilistic critical-group calculation does not.

The International Commission on Radiological Protection (ICRP) endorses calculating the average dose to a homogenous[8] critical group. The group should include the person at highest exposure and risk. ICRP's critical-group concept has been useful in evaluating the safety of operating facilities, where habits of the present population at risk can be included in the analysis of doses and risks.

However, because the habits and population at risk in the far future are not known, ICRP recommends (see "Radiation Protection Principles for the Disposal of Solid Radioactive Waste", ICRP-46 [ICRP, 1985a]):

> "When an actual group cannot be defined, a hypothetical group or representative individual should be considered who, due to location and time, would receive the greatest dose. The habits and characteristics of the group should be based upon present knowledge using cautious, but reasonable, assumptions. For example, the critical group could be the group of people who might live in an area near a repository and whose water would be obtained from a nearby groundwater aquifer. Because the actual doses in the entire population will constitute a distribution for which the critical group represents the extreme, this procedure is intended to ensure that no individual doses are unacceptably high." [emphasis added]

ICRP-43 also endorses the single hypothetical individual when dealing with conditions far in the future:

> "In an extreme case it may be convenient to define the critical group in terms of a single hypothetical individual, for example when dealing with conditions well in the future which cannot be characterized in detail" [ICRP, 1984b]. [Emphasis added.]

---

[8] ICRP recommends that the group include the most exposed individual and that there be no more than a tenfold variation in exposure within the critical group.

On the basis of the above quotes from ICRP, I concur with UK's NRPB and others that the subsistence farmer is the appropriate single hypothetical individual to be considered for dose and risk calculations for the distant future. The diet and dose response of the subsistence farmer are to be based on present knowledge, as recommended by ICRP. It is cautious and reasonable that there can exist in the future a farmer whose food intake is largely that grown in contaminated water. Because the subsistence-farmer calculation is bounding, it represents the extreme of the actual doses in the entire population. Protecting the subsistence farmer will ensure that no individual doses are unacceptably high. [Emphasis shows connection to ICRP-46 and ICRP-43 recommendations.]

Those wishing to identify a critical group can imagine a group that would include the subsistence farmer, subject to ICRP's homogeneity criterion that the dose or risk to individuals within the group should vary no more than tenfold.[9]

The full-time subsistence farmer, who receives no food and water from noncontaminated sources, is obviously the bounding scenario. We assign a probability of unity that he can exist. Some part-time farmers will be included in the data for the Committee's probabilistic analysis, because they exist now in the Amorgosa Valley. However, because the Committee's method is expected to synthesize a continuous probabilistic distribution function of occupancy and exposure to radiation, the full-time subsistence farmer will not be found on that distribution. Speculation that the Committee's probabilistic approach will yield the full-time subsistence farmer as the individual with maximum exposure is not valid. Methods of Appendices C and D do not converge.

---

[9] The Committee makes much of the claim that the probabilistic exposure scenario of Appendix C can predict the dose/risk variation within the calculated critical group, so that the average dose within the group can be calculated. However, the ratio of maximum to average dose/risk must lie between one and ten, if the critical group meets ICRP's homogeneity criterion. An assumed linear variation results in a ratio of two, as assumed in the subsistence-farmer approach. I have already noted that the large uncertainties in calculating geosphere performance, together with the additional uncertainties inherent in the Committee's proposed probabilistic exposure calculations, do not justify such attempts to refine the ratio beyond that assumed above. Again, calculated exposures from the probabilistic scenario are of questionable validity, whereas the subsistence-farmer results are conservative and bounding.

The probabilistic approach can yield a maximum value of the dose/risk calculated by that method. However, that maximum is not the maximum to which future people can be exposed. It is not bounding. Although the probabilistic approach may suffice for those who desire a self-consistent calculational exercise as a matter of policy, it cannot fulfill the desired goal that "if the individual at calculated maximum risk is suitably protected, all other individuals will also be protected."

The Committee justifies its probabilistic scenario on ICRP's use of the words "based upon present knowledge". By attempting to extrapolate data on the present nearby population to predict probabilities of location, number, and exposure of future people, the Committee overextends its use of present knowledge. The Committee's probabilistic approach is neither "cautious" nor "reasonable". It can lead incorrectly to low values of calculated doses and risks to a group selected as "the critical group". The Committee's probabilistic procedure cannot ensure that no individual doses are unacceptably high. It does not fulfill the recommendations of ICRP quoted above. (see Comments 6 and 7).

According to the Committee, probabilities of habits and behavior of future humans can be derived from data on any arbitrarily chosen reference population, whether past, future, hypothetical, or present. The Committee adopts the present population only to illustrate the probabilistic method. However, past, future, or hypothetical reference populations could not provide the kind of "present-knowledge" human data that the Committee claims must be used to satisfy ICRP's recommendation. Therefore, the Committee's definition of reference population does not satisfy the Committee's interpretation of ICRP guidance concerning use of "present knowledge" for establishing a critical group.

The Committee does not claim that its probabilistic exposure scenario can predict the habits of future generations; it only presents what is said to be a self-consistent calculation of individual risks based on assumed extrapolation from an arbitrary reference population. Even if correctly formulated, the Committee's probabilistic approach can tell us nothing about whether a subsistence farmer family can and will exist during any of the thousands of generations when people can be at significant risk. Common sense tells us that it is not reasonable to assume that the probability that a subsistence-farmer will not exist during one of the many thousands of future generations is necessarily low. The subsistence farmer is the bounding scenario for calculating doses and risks to the types of people who, by policy, are to be protected. Therefore, protecting a critical group that includes the subsistence farmer is necessarily the only cautious and reasonable approach that will fulfill ICRP's goal of ensuring that no individual doses are unacceptably high. Clearly, the Committee's less stringent probabilistic approach cannot ensure that no individual doses are unacceptably high.

The Committee wishes to avoid calculating dose/risk to a single individual or to a family of subsistence farmers as adopted by NRPB and USNRC (see Comment 2). The Committee does not explain why. As quoted above, ICRP-46 accepts a "representative individual" for calculation, and ICRP-43 endorses the single hypothetical individual when dealing with conditions far in the future:

The Committee's argument against the subsistence farmer appears in the following statement in Chapter 2 of the Committee's report:

> "... we believe that a reasonable and practicable objective is to protect the vast majority of members of the public while also ensuring that the decision on the acceptability of a repository is not prejudiced by the risks imposed on a very small number of individuals with unusual habits or sensitivities. The situation to be avoided, therefore, is an extreme case defined by unreasonable assumptions regarding the factors affecting dose and risk, while meeting the objectives of protecting the vast majority of the public." [From Chapter 2, emphasis added]

The objectives are laudable, but the Committee and others [EPRI, 1994] infer that it is necessary to calculate doses and risks to groups of future people rather than to an individual such as a subsistence farmer, contradicting ICRP [ICRP 1984,1985].

The Committee infers, in the above quote, that it is the subsistence farmer (or maximally exposed individual) who is to be ruled out because of "unusual habits or sensitivities." The Electric Power Research Institute (EPRI) reaches a similar conclusion and so states. The Committee and EPRI have apparently adopted words by UK's NRPB:

> "The purpose of the critical group concept ....is to ensure that the vast majority of members of the public do not receive unacceptable exposures, whilst at the same time ensuring that

> decisions as to the acceptability or otherwise of a practice are not prejudiced by a very small number of individuals with unusual habits." [Barraclough, *et al.*, 1992]

Both the Committee and EPRI have taken the NRPB words out of context and have misinterpreted NRPB. As is apparent from the full quotes of NRPB (see Comment 2), the individuals with "unusual habits" whom NRPB refers to are those with unusual sensitivities to radiation and with unusual diets.[10] It is a mistake to assume that the NRPB statement about "a very small number of individuals" refers to the subsistence farmer, because NRPB endorses the use of the subsistence farmer.

Because the Committee's probabilistic approach cannot predict the actual habits of future people, and because it will predict lower doses and risks than would be calculated for a subsistence farmer, there will be no way of knowing whether the Committee's objective to protect the vast majority of members of the public will be fulfilled.

5. There is consensus that the subsistence-farmer approach is consistent with the critical-group concept.

The USNRC adopts a critical group that consists of a subsistence-farmer family of three people [McCartin, *et al.*, 1994]. According to Neel [1995] this is the "reference-man" concept developed by ICRP. Neel also states that a similar approach has been taken by a working group within BIOMOVS, the international Biospheric Model Validation Study, for making long term assessments of dose. BIOMOVS is a cooperative effort by selected members of the international nuclear community to develop and test models designed to quantify the transfer and bio-accumulation of radionuclides in the environment.

---

[10] Some precedence for excluding such individuals arises from UK's recent Sizewell Inquiry, which concerned a proposal to construct a new operating facility that could affect existing populations. A study of present population revealed that several individuals subsisted almost entirely on clams obtained in the vicinity. Because of the unusual diet, UK did not include those individuals in its analysis of the critical group.

In speaking of the critical-group concept, USNRC states:

> " ....the specific individuals who may receive the highest exposures and greatest risks in future time cannot be identified. In these circumstances, it is appropriate to **define** a hypothetical critical group (those persons who receive the highest exposures) because this approach avoids the need to forecast future lifestyles, attitudes to risk, and developments in the diagnosis and treatment of disease." [Neel, 1995]

USNRC's hypothetical critical group is the subsistence-farmer family.

In the same sense, UK's NRPB warns that:

> "...site-specific calculations relating to the biosphere and human behavior should not continue beyond about 10,000 years into the future. Beyond that, simple reference models of the biosphere and human behavior should be adopted in order to calculate the risks to hypothetical reference communities." [Barraclough, *et al.*, 1992]

The reference models adopted by NRPB and by the UK Department of Environment [1994] are for a group involving the subsistence farmer defined herein. (see Comment 2)

Representatives of geologic disposal projects in other countries indicate that their subsistence farmer calculations are consistent with ICRP recommendations.

6. The health standard for geologic disposal of high-level waste must provide adequate and reasonable protection of public health, but it must not be so stringent as to preclude practicable disposal.

The Committee is concerned that the subsistence-farmer approach is unnecessarily stringent.[11] It prefers the less stringent calculations of doses and risks based on probabilistic calculations of locations and habits of future people. It prefers the calculation of doses and risks based on the probabilistic exposure scenario of Appendix C. That calculation is clearly less stringent than the calculation of dose and risk to a hypothetical subsistence farmer. This is far more important than trying to justify a dose/risk calculation on one of several different interpretations of what ICRP says about exposure scenarios for the long-term.

In the written record of this study there is abundant information, contributed by knowledgeable scientists, concerning the stringency of calculating doses and risks to subsistence farmers as well as information on possible benefits of the Committee's proposed probabilistic approach. That information bears on several questions relevant to this study. Would compliance with a given dose/risk limit be unreasonably difficult if the doses and risks were calculated for subsistence farmers? Would the more conservative subsistence-farmer calculation ensure greater confidence in the adequacy of health protection? Would it do so at the expense of ruling out the nation's present approach to solving the important problem of

---

[11] The Committee may not have fully understood the assumptions specified for the subsistence-farmer calculation. In Chapter 3, the Committee incorrectly states:

> "...the approach in Appendix D specifies *a priori* that a person will be present at the time and place of highest nuclide concentrations in ground water and will have such habits as to be exposed to the highest concentration of radiation in the environment."

The subsistence farmer cannot be exposed to the highest nuclide concentration in ground water. That concentration exists deep underground. Nuclide concentrations in ground water concentrations are calculated for undisturbed flow. Withdrawing ground water by a well will dilute the radionuclide concentration [McCartin *et al.*, 1994]. Appendix D does not deal with "concentrations of radiation in the environment" (see Footnote 2).

disposing of high-level radioactive waste? Is the untried and less conservative probabilistic approach for calculating habits of future humans justified?

These questions cannot now be answered definitively. I concur with the Committee that no judgment can yet be made about whether the proposed Yucca Mountain repository could meet requirements consistent with the recommended standard regardless of what exposure scenario is adopted. We can, however, learn much from the concerns that are not addressed in the Report.

Preliminary calculations of dose/risk to future people who might live near a repository conceptually similar to that proposed for Yucca Mountain were presented, at the Committee's invitation, by DOE contractors [Andrews *et al.*, 1994; Wilson *et al.*, 1994], by representatives of industrial groups, and by others. For such preliminary calculations very conservative assumptions were made concerning geochemical, hydrological, and engineering features. Doses were calculated for the subsistence-farmer. Some of the reported doses were high enough to indicate the need for better data and more detailed analysis.[12]

The Electric Power Research Institute [EPRI, 1994] and its contractor [Wilems, 1993] recommended incorporating probabilities that reduce the calculated doses/risks. Both recommended probabilities that take into account living patterns not yet included in the exposure calculations. These include the probabilities that future people will not be present full time at their residences, that only a small fraction of their food will be contaminated with radioactivity, etc. To illustrate, Wilems

---

[12] The Committee states in Chapter 2:

> "...it is possible to construct scenarios in which an individual could receive a very high dose of radiation, even though only one or two people might ever receive such doses."

The Committee's statement does not properly reflect the studies presented to the committee. Some calculated doses to subsistence farmers were high. The studies made no attempt to estimate the number of subsistence farmers who might receive these doses. The Committee seems to disagree with ICRP recommendations [ICRP, 1984, 1985] that even a single hypothetical individual could replace the critical group for dose calculations when the future population is unknown. (See Comment 4)

assumed probability values and calculated much lower dose/risks, thereby making it easier to meet a given dose/risk limit. In like manner, the Committee's probabilistic exposure approach will predict much lower doses and risks than those calculated for subsistence farmers. (see Comment 7).

The Committee proposes to derive probabilities for future populations from data on living habits of an arbitrarily selected reference population. Many arbitrary assumptions are required. The main effect is to predict lower doses and risks, as was illustrated by Wilems. The Committee's probabilistic approach will clearly be less stringent than the subsistence-farmer approach used in the dose/risk calculations by DOE [Andrews, 1994; Wilson, 1994].

Developing probabilities for any future population that might live in the vicinity of Yucca Mountain is problematical. If the selected reference population is the current population of Amorgosa Valley, these people live 20 or more miles away. The probabilities for future people are to be extrapolated from a study of this reference population. Future people who live closer to Yucca Mountain will have to dig deeper wells. The Committee proposes using existing well data to calculate the probability of finding and extracting ground water nearer Yucca Mountain.

The Committee believes that the extent of nonarable land near Yucca Mountain can lead to lower expected probabilities that future individuals would use underlying contaminated ground water. On the other hand, well water is frequently withdrawn and transported for farming at other locations. Already there is a proposal to extract ground water in the vicinity of Yucca Mountain for commercial use. Where farms are located is not important; where the contaminated ground water is withdrawn is important.[13] The Committee's conclusion that future inhabitants will be at no risk if not living over contaminated ground water [Appendix C] is not defensible and is one of the many unjustified assumptions that will reduce the calculated dose/risk.

---

[13] The Committee does indicate [Appendix C] that other sources of water should be considered for areas <u>outside</u> the calculated plume of contaminated ground water. The calculated plume will be very broad, however. The Committee gives no recognition to the importance of considering transport of contaminated well water to farmers who will live within the projected area of the underground plume.

Arbitrary assumptions could result in low probabilities of exposure or to a conclusion that a less stringent calculation of doses and risks is warranted. For example, one such assumption is that the future population could be large in number but confined to present population boundaries, effectively imposing a 20-mile exclusion distance. Another such assumption is that, if not confined to present boundaries, future populations would use wells no deeper than used by the present population 20 or more miles away, so future people nearer the repository would have to import food and water produced farther from Yucca Mountain. Such assumptions would certainly result in low probabilities and lower calculated doses and risks. The assumptions are arbitrary and not defensible.

One might argue that the benefits of the arid climate and present low population near Yucca Mountain will be lost if doses and risks are calculated for individuals exposed to radioactivity extracted from wells. However, there are advantages and disadvantages. The arid climate and lack of flowing surface water may invite people to use water extracted from wells. At other sites flowing surface water may dilute the contaminated ground water before it is used by humans [NRC, 1983]. However, at least two projects in other countries are calculating doses/risks to subsistence farmers who are assumed to use contaminated ground water directly, similar to what would occur at Yucca Mountain. These projects expect that they can meet performance goals similar to those suggested in this study.

There is no evidence that would justify adopting a calculational method for Yucca Mountain compliance that is less stringent than the subsistence-farmer method adopted in other countries. The recent individual dose/risk calculations for the proposed Yucca Mountain repository are preliminary. They involve many conservative and unrealistic assumptions about engineering features. The hydrogeological, environmental, and engineering-design features of Yucca Mountain do not suggest that a less stringent calculational approach is necessary. Indeed, there are many features that can favor long-term performance.[14]

---

[14] A repository in unsaturated tuff at Yucca Mountain may have much greater dilution of many radionuclides than repositories in those other countries that calculate doses from using ground water contaminated by waste buried in saturated rock. For radionuclides whose release from waste solids is limited by

(continued...)

If a less stringent approach were justified, it would be far better to adopt a less restrictive value of the dose/risk limit than to adopt a probabilistic exposure calculation that will be so difficult to defend. The probabilistic exposure approach is neither cautious nor reasonable. It cannot ensure that no future individual will receive an unacceptable dose or risk.

7. Calculational techniques described in Appendix C are not mathematically valid. They can be manipulated to produce even lower calculated doses/risks.

The Committee proposes to establish full distributions, with respect to space and time, of numbers of future populations and of their water and food sources in the area surrounding Yucca Mountain. The surrounding area is to be divided into subareas. Each subarea can be arbitrarily large and can contain as many people as one chooses. Based on the assumed and extrapolated probabilities of location and living habits of future people, and using calculated concentrations of contaminants in ground water, doses and risks to individuals in each subarea are to be calculated.[15] The arithmetic average of all individual doses/risks in a

---

(...continued)

solubility, the release rate from the solid waste will be far less for the unsaturated repository, because of the low infiltration rate of ground water in the unsaturated zone. Contaminants in this infiltration flow will be highly diluted when they reach the underlying aquifer. Water flow past waste packages in saturated rock will be far greater, as will the release rate of such radionuclides to ground water. It would be premature to conclude that Yucca Mountain would be at a disadvantage relative to other repositories. There is no basis for proposing a less-stringent calculation of doses and risks for Yucca Mountain.

[15] The Committee's probabilistic method will yield calculated individual doses and risks that will depend on the population density and number of people in a subarea. The Committee has not explained how the growth in population is to be predicted; how the probabilistic distributions of number of people with respect to location and time, together with probabilistic distributions of parameters of occupancy, food source, etc., can result in a map of potential farm

(continued...)

subarea is to be calculated. The subarea that is calculated to have the highest average dose/risk, together with additional subareas in which the average subarea risk is greater than or equal to one tenth of the risk in the subarea with maximum average risk, is said to define a critical subgroup. The average subgroup risk is said to be calculated as the arithmetic mean of the average risks of the selected subareas. The process is repeated for many different samplings of parameters that affect the probabilistic distributions, to produce new values of the critical-subgroup risks. The critical-group risk is said to be the arithmetic average of all calculated critical-subgroup risks. (see Appendix C)

However, the Committee's interpretation of ICRP would require calculating doses/risks for individuals over a large area, properly utilizing the many probability distribution functions of the geosphere and biosphere to calculate probabilistic distributions and expected values of consequences, selecting the individuals whose risks are within the top ten percent, and calculating the average risk of that critical group. This method is mathematically inconsistent with the Committee's proposed subarea/subgroup method. It would be fortuitous if the two methods were to produce the same result. The subarea method will tend to calculate lower doses and risks.

The Committee's subarea method will not necessarily yield a critical group that includes the individual at maximum exposure and risk. That individual may be located in a subarea wherein are many individuals at much lower exposure. The subarea size and boundaries are arbitrary. There could result so low an arithmetic average dose for that entire subarea that it would not be selected for calculating the critical group. The

---

(...continued)

density or water use; how many such maps will have to be generated and how they are to be used in conjunction with the many equivalent maps of sampled plume concentration; how population changes from the many expected cycles of climate change are to be calculated; how the expected values of consequence to individuals at various times and locations are to be obtained without simultaneously sampling distribution functions of geosphere performance and biosphere performance; and how the probability distribution functions are to be generated if any of the other arbitrary reference populations suggested by the Committee are adopted.

resulting "critical group" would not meet the ICRP criterion that the individual of greatest exposure should be included.

Further, to achieve a lower calculated average dose in a subarea, one would need only to move the outer boundaries of the subarea farther from Yucca Mountain, to add more people exposed to lower doses. Applied to all subareas, arithmetic average doses would decrease, as would the average dose for the calculated "critical group." The repository would appear to be safer! The calculated critical-group doses and risks would be much lower than those for a critical group that includes a subsistence farmer. Or, to lower the calculated risk, a different reference population could be selected. The calculated lower doses and risks would be obtained with an illusion of safety, but with a serious loss of credibility.

8. Calculated uncertainties in terms of confidence levels should be used to test compliance.

Large uncertainties are inherent in predictions of the transport of radionuclides to the environment far into the future. Even larger uncertainties would be introduced by the probabilistic approach based on current-population data. The Committee does not discuss how information on uncertainties is to be conveyed and used in compliance determinations.

The performance measure of risk recommended by the Committee is the expected value of the probabilistic distribution of consequences. The Committee recommends that the expected value be compared directly to the risk limit to determine compliance. However, uncertainty should be considered in determining compliance. The expected value (or mean value) conveys nothing about uncertainty. Basing compliance on the expected-value comparison is not sufficient.

A technique commonly used to convey uncertainty is to express the "confidence range" of the result. UK's NRPB illustrates presentation of the results in terms of the 95 percent confidence level. This states a range of values of dose or risk, such that 95 percent of the possible values of the distribution are calculated to fall within that range. NRPB then compares that range with a dose or risk limit [Barraclough *et al.*, 1992] . Effectively, the upper value of the range becomes the dose or risk value for determining compliance. Methods of calculating confidence levels are well documented.

Presenting 90 or 95 percent confidence levels is done extensively for the geologic disposal projects in Sweden and Finland. It is a technique commonly used in the U.S., particularly when the results are important to public understanding and acceptance [e.g., Farris *et al.*, 1994ab].

9. The Yucca Mountain project needs a soundly based standard for performance assessment and compliance. The U.S. program needs to share the benefits of an international approach towards developing standards and technology for geologic disposal.

A standard and regulatory guidance to ensure public health and safety in the long-term for geologic disposal must include both a regulatory limit as well as guidance on assumptions of habits of future individuals and population groups to be adopted in calculating those individual doses and risks. I agree with and support the Committee's recommendation that the measure of performance best suited to assure public health and safety for the long term is the dose and risk to future individuals. That measure was adopted by the National Research Council's Waste Isolation Systems Committee (WISP) [NRC, 1983], after review and analysis of the release limits then proposed by EPA, and was subsequently incorporated in EPA's standard, 40 CFR 191. The WISP Panel concluded that individual dose is a traditional and sound measure in assessing public-health protection.[16] It was also noted that most, or possibly all, other countries undertaking geologic disposal use individual dose (or individual risk) as a performance measure. Adopting the same performance measure as other countries would provide a framework for interchanging and sharing information with other countries on the developing technology for geologic disposal. The technical approach to design and performance analysis, for the purpose of ensuring long-term safety, depends greatly on the performance criterion that is adopted.

---

[16] I agree that individual risk is better than dose as a measure of performance, because it allows for possible future changes in the dose/risk conversion factor. As has already been explained in the Panel's report, calculated values of radiation dose would include probabilistic analysis of uncertainty and probabilities, if calculable, of being exposed to the radiation.

The EPA release-limit standard has now been set aside for Yucca Mountain after considerable effort has been expended in designing for compliance with that standard. Adopting a performance measure based on individual dose and risk is an important step towards developing a standard that has a clear basis for protection of public health. The international consensus favoring individual dose/risk is likely to ensure understanding and support of its adequacy for protecting public health. Both the technical community and the general public can be reasonably expected to see the virtues in individual dose/risk as a performance measure.

However, acceptance of the use of individual dose/risk for ensuring safety cannot be expected if methods of calculating doses and methods of assessing compliance are not visibly sound, suitably conservative and understandable. Selecting an exposure scenario to be used in calculating long-term doses is a crucial step that can greatly affect the magnitude of calculated individual doses and risks. If calculated risks to the bounding subsistence farmer are found be within compliance limits, then no future individual doses would be unacceptably high.[17] In contrast, the probabilistic exposure calculation is too vaguely defined, subject to too many arbitrary and unconservative policy decisions and subject to too many questions of validity to meet any reasonable test of acceptability, once the shortcomings of that approach have been sufficiently understood.

Adopting the probabilistic exposure calculation would again put the U.S. repository program on a course divergent from that in other countries. One must expect continued questioning, by the scientific community, by the public, and by geologic programs in other countries, of why the U.S. finds it necessary to adopt such a unconservative approach to regulating geologic disposal. The U.S. program needs to share the benefits of an international approach towards developing standards and technology for geologic disposal, including how to calculate individual doses and risks for compliance determination.

The U.S. geologic disposal program needs a standard, including regulatory guidance, that can be clearly implemented and that can be expected to survive challenges. Serious challenges are likely to arise many years hence when an application is finally submitted to the regulatory agency for licensing determination. By that time an enormous investment of public and electric-utility funds will have been expended in the

---

[17] See Comment 4.

development of repository technology and in the performance analysis to assure compliance with the new performance standard. Of the total funds expended, most will have been to develop technological and geosphere information, to produce designs of engineering barriers that can assure safety, to produce calculations of individual risk for determining compliance, and for administration and services. The cost of constructing the repository is expected to be small in comparison. Therefore, it is essential that the new regulatory standard and guidance be on firm ground so that this enormous effort, measured in money and time, is not wasted. Adopting an individual dose/risk standard is a step in that direction. Adopting the probabilistic exposure calculation, however, would leave the U.S. program vulnerable to future challenge on grounds of reasonable assurance of safety.

I advocate an approach that ensures that all individuals are suitably protected, that is based on sound science and logic, and that does not compromise scientific validity and credibility under the aegis of policy.

Adopting the unconservative probabilistic exposure scenario will undermine public confidence. The scientific community and the public will find it difficult to understand why the Committee endorses the probabilistic exposure scenario that is demonstrably less stringent in protecting public health than the subsistence-farmer approach, the approach that has been adopted for geologic disposal projects in other countries and in the U.S.

REFERENCES FOR APPENDIX E

Andrews, R. W., T. F. Dale, and J. A. McNeish, "Total System Performance Assessment -- An Evaluation of the Potential Yucca Mountain Repository," Yucca Mountain Site Characterization Project, INTERA, Inc., Las Vegas, Nevada, 1994.

Barraclough, I. M., S. F. Mobbs, and J. R. Cooper, "Radiological Protection Objectives for the Land-Based Disposal of Solid Radioactive Wastes," Documents of the National Radiation Protection Board (NRPB), 3, No. 3, 1992.

Charles, D., and G. M. Smith, "Project 90 Conversion of Releases From the Geosphere to Estimates of Individual Doses to Man," Swedish Nuclear Regulatory Commission, SKI Technical Report 91:14, 1991.

Davis, P. A., R. Zach, M. E. Stephens, B. D. Amiro, G. A. Bird, J. A. K. Reid, M. T. Sheppard, S. C. Sheppard, M. Stephenson, "The Disposal of Canada's Nuclear Fuel Waste: The Biosphere Model, BIOTRAC, for Postclosure Assessment," AECL-10720, 1992.

Electric Power Research Institute (EPRI), "A Proposed Public Health and Safety Standard for Yucca Mountain: Presentation and Supporting Analysis" Report EPRI TR-104012, April 1994.

Farris, W. T., B. A. Napier, T. A. Ikenberry, J. C. Simpson, D. B. Shipler, "Atmospheric Pathway Report, 1944-1992," Hanford Environmental Dose Reconstruction Project, Pacific Northwest Laboratories, PNWD-2228 HEDR, 1994a.

Farris, W. T., B. A. Napier, T. A. Ikenberry, J. C. Simpson, D. B. Shipler, "Columbia River Pathway Dosimetry Report," 1944-1992," Hanford Environmental Dose Reconstruction Project, Pacific Northwest Laboratories, PNWD-2227 HEDR, 1994b.

International Commission on Radiological Protection (ICRP), "Radiation Protection Principles for the Disposal of Solid Radioactive Waste," Report ICRP-46, Annals of the ICRP, 1985a.

International Commission on Radiological Protection (ICRP), "Principles of Monitoring for the Radiation Protection of the Population," Report ICRP-43, Annals of the ICRP, 1985b.

International Commission on Radiological Protection (ICRP), "1990 Recommendations of the International Commission on Radiological Protection," Report ICRP-60, Annals of the ICRP, Pergamon, 1991.

Leigh, C. D. *et al.*, "User's Guide for GENII-S: A Code for Statistical and Deterministic Simulation of Radiation Doses to Humans from Radionuclides in the Environment," Sandia National Laboratories, SAND-91-0561, 1993.

McCartin, T., R. Codell, R. Neel, W. Ford, R. Wescott, J. Bradbury, B. Sagar, J. Walton, "Models for Source Term, Flow, Transport and Dose Assessment in NRC's Iterative Performance Assessment, Phase 2," Proc. International Conference on High Level Radioactive Waste Management, Las Vegas, NV, 1994.

Napier, B. A., R. A. Peloquin, D. L. Streng, J.V. and J.V. Ramschell, "GENII: The Hanford Environmental Radiation Dosimetry Software System," Richland, Washington, Pacific Northwest Laboratory, PNL-6584, 1988.

Neel, R. B., "Dose Assessment Module", in NRC Iterative Performance Assessment Phase 2: Development of Capabilities for Review of A Performance Assessment for a High-Level Waste Repository," R. G. Wescott, M. P. Lee, T. J. McCartin, N. A. Eisenberg, and R. B. Baca, eds., U. S. Nuclear Regulatory Commission, NUREG-1464, 1995.

Pigford, T. H., J. O. Blomeke, T. L. Brekke, G. A. Cowan, W. E. Falconer, N. J. Grant, J. R. Johnson, J. M. Matuszek, R. R. Parizek, R. L. Pigford, D. E. White, "A study of the Isolation System for Geologic Disposal of Radioactive Wastes,", National Academy Press, Washington, D.C., 1983.

Planning Information Corporation, Nye County, Nevada, "Socioeconomic Conditions and Trends," 1993.

Rechard, Rob P., Ed., "Performance Assessment of the Disposal in Unsaturated Tuff of Spent Nuclear Fuel and High-Level Waste Owned by U.S. Department of Energy, Vol. 1: Methodology and Results," Sandia National Laboratories, SAND94-2563/2, March 1995.

Switzerland National Cooperative for the Storage of Radioactive Waste, "Nuclear Waste Management in Switzerland: Feasibility Studies and Safety Analyses," Project Report NGB 85-09, June 1985.

Switzerland National Cooperative for the Storage of Radioactive Waste, "Kristallin - I: Safety Assessment Report," Technical Report 93-22E, February 1994.

UK Department of the Environment, "Review of Radioactive Waste Management Policy: Preliminary Conclusions," August, 1994.

van Dorp, F., NAGRA Project Manager for Biosphere Models, Switzerland, Private Communication, 1994.

Vieno, T., A. Hautojarvi, L. Koskinen, and H. Nordman, "TVO-92 Safety Analysis of Spent Fuel Disposal," Report YJT-92-33E, Technical Research Centre of Finland, Nuclear Engineering Laboratory, Helsinki, 1992.

Wilems, R. E., Presentation to the National Academy of Sciences Committee on Technical Bases for Yucca Mountain Standards, 1993.

Wilson, M.L., J. H. Gautahier, R. W. Barnard, G. E. Barr, H. A. Dockery, E. Dunn, R. R. Eaton, D. C. Guerin, N. Lu, M. J. Martinez, R. Nilson, C. A. Rautman, T. H. Robey, B. Ross, E. E. Ryder, A. R. Schenker, S. A. Shannon, L. H. Skinner, W. G. Halsey, J. D. Gansemer, L. C. Lewis, A. D. Lamont, I. R. Triay, A. Meijer, D. E. Morris, "Total System Performance Assessment for Yucca Mountain," Sandia Laboratories, SAND93-2675, 1994.

APPENDIX F

THE COMMITTEE CHAIR'S PERSPECTIVE ON APPENDIX E

ROBERT W. FRI

In Appendixes C and D, we have presented alternative approaches that EPA might wish to consider in selecting an exposure scenario to be used in calculating compliance with the standards. As noted in Chapter 3 of the report, these approaches differ chiefly in the assumptions and calculational methods used in estimating the exposure of future persons who might be near the repository site. However, there is little scientific basis for predicting events far into the future, such as where people will live, and so developing an exposure scenario for testing repository compliance with the standards is inherently a policy choice.

Throughout our report, we have avoided making recommendations that involve policy choices on the grounds that there is by definition a limited scientific basis for selecting one policy alternative over another. We have instead tried to use available technical information and judgment to suggest a starting point for the rulemaking process that will lead to a policy decision. As noted in Chapter 3, a majority of the committee considers the approach of Appendix C to be more clearly consistent with the technical criteria that define the critical group in the exposure scenario, and therefore believes that EPA should propose an approach along the lines of Appendix C. The committee recognizes, however, that other approaches might meet these criteria.

I believe that, in his personal statement, Dr. Pigford has become an advocate for a particular choice. He clearly prefers the approach of Appendix D and presents arguments both for his position and against the alternative. He is of course entitled to make this argument. It is important, however, to understand that the argument being presented is fundamentally a policy argument rather than a scientific one.

Nevertheless, the issue raised here is an important one. Dr. Pigford advocates an assumption that results, in his words, in calculating ". . .the extreme of the actual doses in the entire population". In contrast,

Chapter 2 of the report adopts the basic principle of the International Commission for Radiological Protection that the standard should avoid "...an extreme case defined by unreasonable assumptions regarding factors affecting dose and risk". Although Appendix D and Dr. Pigford postulate a subsistence-farmer scenario based on cautious, but reasonable, assumptions (as described in Chapter 2), some members of the committee believe that the approach advocated by Dr. Pigford could become just such an extreme case.

Determining when the assumptions in an exposure scenario pass from cautious to extreme is thus a crucial issue in the rulemaking process. As such, it requires the fullest and most open public discussion.

GLOSSARY

Accessible environment
: Those portions of the environment directly in contact with or readily available for use by human beings. Includes the earth's atmosphere, the land surface, aquifers, surface waters, and the oceans. In 40 CFR 191, the environment outside a surface defined as enclosing a controlled area.

ALARA
: An acronym for "as low as reasonably achievable", a concept meaning that the design and use of sources, and the practices associated therewith, should be such as to ensure the exposures are kept as low as is reasonably practicable, economic and social factors being taken into account.

Backfill
: The material used to refill the excavated potions of a repository or of a borehole after waste has been emplaced.

Becquerel
: International unit of radioactivity. Symbol Bq = 1 disintegration per second.

Biosphere
: The region of the earth in which environmental pathways for transfer of radionuclides to living organisms are located and by which radionuclides in air, ground water, and soil can reach humans to be inhaled, ingested, or absorbed through skin. Humans can also be exposed to direct irradiation from radionuclides in the environment.

Borehole
: A cylindrical excavation in the earth, made by a rotary drilling device.

Canister A closed or sealed container for nuclear fuel or other radioactive material, which isolates and contains the contents; it might rely on other containers (e.g. a cask) for shielding.

Collective dose The sum of the individual doses received in a given period of time by a specified population from exposure to a specified source of radiation.

Critical group Originally defined for dose by the ICRP (ICRP, 1977, p.17; ICRP, 1985b, pp.3-4) as a relatively homogeneous group of people whose location and habits are such that they are representative of those individuals expected to receive the highest doses as a result of the discharges of radionuclides. The definition is extended to risk in Chapter 2 of this report.

Critical pathway The dominant environmental pathway through which a given radionuclide reaches the critical group.

Disposal Permanent isolation of spent nuclear fuel or radioactive waste from the accessible environment with no intent of recovery, whether or not such isolation permits the recovery of such fuel or waste.

Disposal package The primary container that holds, and is in contact with, solidified high-level radioactive waste, spent nuclear fuel, or other radioactive materials, and any overpacks that are emplaced at a repository.

Dose A measure of the radiation received or absorbed by a target.

Dose rate Absorbed dose per unit time.

Engineered barrier system The waste form, cladding, backfill, and canister, all of which are intended to retard disperson of radionuclides.

Exposure Irradiation of persons or materials. Exposure of persons to ionizing radiation can be either:

1. external exposure, irradiation by sources outside the body; or
2. internal exposure, irradiation by sources inside the body.

Fault A surface or zone of rock fracture along which there has been displacement.

Geologic repository A system that is intended to be used for, or might be used for, the disposal of radioactive wastes in excavated geologic media. A geologic repository includes: (1) the geologic repository operations area and (2) the portion of the geologic setting that provides isolation of the radioactive waste.

Ground water Water that permeates the rock strata of the Earth, filling their pores, fissures and cavities. (It excludes water of hydration.)

Ground water transport The principal means by which radionuclides can be mobilized from an underground repository and moved into the biosphere. Avoiding or minimizing such transport is the basis for selecting and designing repository systems.

Half-life In physics, the time required for the transformation of one-half of the atoms in a given radioactive decay process, following the exponential law (physical half-life).

High-level radioactive waste The highly radioactive material resulting from the reprocessing of spent nuclear fuel, including liquid waste produced directly in reprocessing and any solid material derived from such liquid waste that contains fission products in sufficient concentrations. Other highly radioactive material that the U.S. Nuclear Regulatory Commission, consistent with existing law, determines by rule requires permanent isolation. Also referred to as high-level waste (HLW).

IAEA International Atomic Energy Agency is an autonomous intergovernmental organization established by the United Nations. It is authorized to foster research and development in the peaceful uses of nuclear energy, to establish or administer health and safety standards, and to apply safeguards in accordance with the Treaty of the Non-Proliferation of Nuclear Weapons.

ICRP The International Commission on Radiological Protection is an international organization that develops guidance and standards for radiological measurement and protection of public and occupational health. The ICRP is composed of a Chairman and never more than 12 other members. The selection of the members is made by the ICRP from nominations submitted to it by the National Delegations to the International Congress of Radiology and the ICRP staff itself. Members of the ICRP are chosen on the basis of their recognized activity in the fields of medical radiology, radiation protection, physics, biology, genetics, biochemistry, and biophysics. The ICRP's rules require that its members be elected every four years.

Linear model Also, linear dose-effect relationship; expresses the health effect, such as mutation or cancer as a direct (linear) function of dose.

Natural background radiation The amount of radiation to which a member of the population is exposed from natural sources, such as terrestrial radiation due to naturally occurring radionuclides in the soil, cosmic radiation originating in outer space, and naturally occurring radionuclides deposited in the human body.

NCRP National Council on Radiation Protection and Measurements is an organization of nationally recognized scientists who share the belief that significant advances in radiation protection and measurement can be achieved through cooperative effort. It conducts research focusing on safe occupational exposure levels and disseminates information.

Performance assessment — Analysis to predict the performance of the system or subsystem, followed by comparison of the results of such analysis with appropriate standards or criteria.

Population dose — The sum of the doses to all the individuals in a specified group. In units of person-sievert or person-rem. (Also called collective dose.)

Radioactive decay — The spontaneous transformation of a nuclide into one or more different nuclides accompanied by either the emission of energy or particles. Unstable atoms decay into a more stable state, eventually reaching a form that does not decay further or is very long-lived.

Radioactive waste — Any material that contains or is contaminated with radionuclides at concentrations or radioactivity levels greater than the exempt quantities established by the competent authorities and for which no use is foreseen.

Radionuclide — A radioactive species of an atom characterized by the constitution of its nucleus.

Rem — A unit of dose equivalent to one-hundredth of a sievert (1 cSv).

Repository	Any system licensed by the U.S. Nuclear Regulatory Commission that is intended to be used for, or can be used for, the permanent deep geologic disposal of high-level radioactive waste and spent nuclear fuel, whether or not such system is designed to permit the recovery, for a limited period during initial operation, of any material placed in such system. Such term includes both surface and subsurface areas at which high-level radioactive waste and spent nuclear fuel handling activities are conducted.
Risk	In the context of this study, risk is the probability of an individual receiving an adverse health effect and includes the probability of getting a dose.
Saturated zone	That part of the earth's crust beneath the regional water table in which all voids, large and small, are ideally filled with water under pressure greater than atmospheric.
Seismic	Pertaining to, characteristic of, or produced by earthquakes or earth vibrations.
Sievert	International Unit (SI) of equivalent radiation dose. The product of the absorbed dose and the quality factor of the radiation. Symbol Sv.
Spent fuel	Fuel that has been withdrawn from a nuclear reactor following irradiation, the constituent elements of which have not been separated by reprocessing.

Stochastic health effects Random events leading to effects whose probability of occurrence in an exposed population (rather than severity in an affected individual) is a direct function of dose; these effects are commonly regarded as having no threshold; hereditary effects are regarded as being stochastic; some somatic effects, especially carcinogenesis, are regarded as being stochastic.

Storage Retention of high-level radioactive waste, spent nuclear fuel, or transuranic waste with the intent to recover such waste or fuel for subsequent use, processing, or disposal.

Tuff Rock formed from consolidated volcanic ash.

Units

Units[a]	Symbol	Conversion Factors
Becquerel (SI)	Bq	1 disintegration/sec = 2.7×10^{-11} Curies
Curie	Ci	3.7×10^{10} disintegrations/sec = 3.7×10^{10} Becquerels
Gray (SI)	Gy	1 Joule/kg = 100 rads
Rad	rad	100 ergs/gram = 0.01 Grays
Rem	rem	0.01 Sievert
Sievert (SI)	Sv	100 rems

[a] International Units are designated SI.

Unsaturated zone The zone between the land surface and the regional water table. Generally, fluid pressure in this zone is less than atmospheric pressure, and some of the voids might contain air or other gases at atmospheric pressure. Beneath flooded areas or in perched water bodies the fluid pressure locally may be greater than atmospheric. Also referred to as vadose zone.

Vadose zone See definition for unsaturated zone.

Volcanism The process by which magma and the associated gases rise into the crust and are extruded onto the earth's surface and into the atmosphere.

Waste form The radioactive waste materials and any encapsulating or stabilizing matrix.

Waste package The waste form and any containers, shielding, packing and other absorbent materials immediately surrounding an individual waste container.

Water table The upper surface of the saturated zone on which the water pressure in the porous medium equals atmospheric pressure.

REFERENCES

Andrews, R.W., T.F. Dale, and J.A. McNeish. 1994. Total System Performance Assessment — An Evaluation of the Potential Yucca Mountain Repository. Prepared for U.S. Department of Energy's Yucca Mountain Site Characterization Project. INTERA, Inc., Las Vegas, Nev.

Barraclough, I.M., S.F. Mobbs, and J.R. Cooper. 1992. Radiological Protection Objectives for the Land-Based Disposal of Solid Radioactive Wastes. Documents of the National Radiation Protection Board (NRPB), 3, No. 3, London, U.K.

Berkovitz, D.M. 1992. Pariahs and prophets: nuclear energy, global warming, and intergenerational justice. Columbia J. Environ. Law 17:245.

Charles, D., and G.M. Smith. 1991. Project 90 Conversion of Releases From the Geosphere to Estimates of Individual Doses to Man. Swedish Nuclear Regulatory Commission. SKI Tech. Rep. 91:14, July.

Comar, C. 1979. Risk: A pragmatic de minimis approach. Science 2-3:319.

Crowe, B.M., F.V. Perry, G.A. Valentine, P.C. Wallmann, and R. Kossik. 1994. Simulation Modeling of the Probability of Magmatic Disruption of the Potential Yucca Mountain Site. Prepared for the U.S. Department of Energy's Yucca Mountain Site Characterization Project. Los Alamos National Laboratory, NM.

Dansgaard, W., S.J. Johnsen, H.B. Clausen, D. Dahl-Jensen, N.S. Gundestrup, C. U. Hammer, C.J. Hvidberg, J.P. Steffensen, A.E. Sveinbjörnsdottir, J. Jouzel, and G. Bond. 1993. Evidence for general instability of past climate from a 250-kyr ice-core record. Nature 364:218-220.

Davis, P.A., R. Zach, M.E. Stephens, B.D. Amiro, G.A. Bird, J.A.K. Reid, M.T. Sheppard, S.C. Sheppard, and M. Stephenson. 1992. The Disposal of Canada's Nuclear Fuel Waste: The Biosphere Model, BIOTRAC, for Postclosure Assessment. AECL-10720. Atomic Energy of Canada, Ltd.

Dejonghe, P. 1993. Technical Bases for Yucca Mountain Standards (TYMS): Approaches to Health-Based Standards in different countries. Report to the NRC Committee on Technical Bases for Yucca Mountain Standards (TYMS).

DOE (U.S. Department of Energy). 1988. Site Characterization Plan: Yucca Mountain Site, Nevada Research and Development Area, Nevada. DOE/RW-0199. Office of Civilian Radioactive Waste Management, U.S. Department of Energy, Washington, D.C.

DOE (U.S. Department of Energy). 1992. Technical Assistance to EPA on 40 CFR 191. Office of Environment, Health, and Safety, U.S. Department of Energy, Washington, D.C.

DOE (U.S. Department of Energy). 1993a. Environmental Restoration and Waste Management Five-Year Plan. Volume I. Washington, D.C. U.S. Government Printing Office.

DOE (U.S. Department of Energy). 1993b. Evaluation of the Potentially Adverse Condition "Evidence of Extreme Erosion During the Quatinary Period" at Yucca Mountain. Topical Report YMP/92-41-TPR. U.S. Department of Energy, Washington, D.C.

Eisenbud, M. 1981. The concept of de minimis dose. Pp 64-75 In Quantitative Risk in Standards Setting. Proceedings No. 2, Proceedings of the Sixteenth Annual Meeting of the National Council on Radiation Protection and Measurements (NCRP), April 2-3, 1980, Bethesda, Md.

EPA (U.S. Environmental Protection Agency). 1991. Human Health Evaluation Manual, Part B. Development of risk-based preliminary remediation goals. Office of Solid Waste and Emergency Response Directive 9285.7-018. U.S. Environmental Protection Agency, Washington, D.C.

EPA (U.S. Environmental Protection Agency). 1992. Summary of EPA Office of Radiation Programs Carbon-14 Dosimetry as Used in the Analysis for High-Level and Transuranic Wastes. U.S. Environmental Protection Agency, Washington, D.C.

EPA (U.S. Environmental Protection Agency). 1993. High-Level and Transuranic Radioactive Wastes. Background Information Document for Amendments to 40 CFR Part 191. EPA 402-R-93-073. U.S. Environmental Protection Agency, Washington, D.C.

EPRI (Electric Power Research Institute). 1994. A Proposed Public Health and Safety Standard for Yucca Mountain. Presentation and Supporting Analysis. EPRI TR-104012. Electric Power Research Institute, Palo Alto, CA.

Henley, E.J., and H. Kumamoto. 1992. Probabilistic risk assessment: Reliability, engineering, design, and analysis. P. 568 in Inst. Elect. Electron Eng. Press. New York.

Holdren, J. 1992. Radioactive Waste Management in the United States: Evolving Prospects. Annual Review of Energy 17:235-259.

IAEA (International Atomic Energy Agency). 1989. Safety Principles and Technical Criteria for the Underground Disposal of High Level Radioactive Wastes. Safety Series 99. IAEA Vienna, Austria.

ICRP (International Commission on Radiological Protection). 1991. Radiation Protection, 1990 Recommendations of the International Commission on Radiological Protection. Pergamon Press, ICRP Publ. 60. Annals of the ICRP. Oxford, U.K.

ICRP (International Commission on Radiological Protection). 1985a. Radiation Protection Principles for the Disposal of Solid Radioactive Waste, ICRP Publ. 46. Annals of the ICRP, Vol 15, No. 4. Pergamon Press, Oxford, U.K.

ICRP (International Commission on Radiological Protection). 1985b. Principles of Monitoring for the Radiation Protection of the Population, ICRP Publ. 43. Annals of the ICRP, Vol. 15, No. 1. Pergamon Press, Oxford, U.K.

ICRP (International Commission on Radiological Protection). 1977. Recommendations of the ICRP. ICRP Publ. 26. Annals of the ICRP, Vol. 1, No. 3. Reprinted (with additions) in 1987. Superseded by ICRP Publ. 60. 1990 (International Commission on Radiological Protection). Pergamon Press, Oxford, U.K.

Jannik, N.O., F.M. Phillips, G.I. Smith, and D. Elmore. 1991. A ^{36}CI chronology for lacustrine sedimentation in the Pleistocene Owens River System: Geol. Soc. Am. Bull. 103:1146-1159.

Napier, B.A., R.A. Peloquin, D.L. Streng, and J.V. Ramschell. 1988. GENI: The Hanford Environmental Radiation Dosimetry Software System. PNL-6584. Pacific Northwest Laboratory. Richland, Washington.

NCRP (National Council on Radiation Protection and Measurements). 1987a. Recommendations on Limits for Exposure to Ionizing Radiation, NCRP Report No. 91. National Council on Radiation Protection and Measurements, Bethesda, Md.

NCRP (National Council on Radiation Protection and Measurements). 1987b. Ionizing Radiation Exposure of the Population of the United States, NCRP Report No. 93. National Council on Radiation Protection and Measurements. Bethesda, Md.

NCRP (National Council on Radiation Protection and Measurements). 1993. Limitation of Exposure to Ionizing Radiation, NCRP Report.

116. National Council on Radiation Protection and Measurements. Bethesda, Md.

NRC (National Research Council). 1957. The Disposal of Radioactive Waste on Land. Washington, D.C. National Academy Press.

NRC (National Research Council). 1983. A Study of the Isolation System for Geologic Disposal of Radioactive Wastes. Washington, D.C. National Academy Press.

NRC (National Research Council). 1990a. Health Effects of Exposure to Low Levels of Ionizing Radiation. BEIR V Report. Washington, D.C. National Academy Press.

NRC (National Research Council). 1990b. Rethinking High-Level Radioactive Waste Disposal. Washington, D.C. National Academy Press.

NRC (National Research Council). 1992. Ground Water at Yucca Mountain: How High Can It Rise?: Washington, D.C. National Academy Press.

NRC (National Research Council). 1994. Science and Judgment in Risk Assessment. Washington, D.C. National Academy Press.

Nitao, J.J., T.A. Busheck, and D.A. Chesnut. 1993. The implications of episodic nonequilibrium fracture-matrix flow on repository performance. Nuclear Technol, Vol. 104, No. 3: 385-402.

Nygaard, O.F., S.L. Brown, K.H. Clifton, J.E. Martin, G.M. Matanoski, H.R. Meyer, R.G. Sextro, P.G. Voilleque, and J.E. Watson. 1993. An SAB Report: Review of Gaseous Release of Carbon-14. EPA-SAB-RAC-93-010. Radiation Advisory Committee of the Science Advisory Board, U.S. Environmental Protection Agency, Washington, D.C.

Okrent, D. 1994. On Intergenerational Equity and Policies to Guide the Regulation of Disposal of Wastes Posing Very Long Term Risks.

UCLA-ENG-22-94. School of Engineering and Applied Science, University of California, Los Angeles.

Ortiz, T.S., R.L. Williams, F.B. Nimick, B.C. Whittet, and D.L. South. 1985. A Three Dimensional Model of Reference Thermal/Mechanical and Hydrological Stratigraphy at Yucca Mountain, Southern Nevada. SAND84-1076. Sandia National Laboratories, Albuquerque, NM.

Reeves, M., N.A. Baker, and J.D. Dupgid. 1994. Review and Selection of Unsaturated Flow Models. Prepared for the U.S. Department of Energy, Yucca Mountain Site Characterization Project. INTERA, Inc., Las Vegas, Nev.

Schiager, K.J., W.J. Bair, M.W. Carter, A.P. Hull, and J.E. Till. 1986. De Minimis Environmental Radiation Levels: Concepts and Consequences. Special Report, Health Physics, Vol. 50, No. 5. Pergamon Press, Oxford, U.K.

Smith, G.M. and D.P. Hodgkinson. 1988. Briefing Document on Alternative Criteria for Disposal of Radioactive Waste in Deep Geological Repositories. Prepared for the Swedish Nuclear Power Inspectorate. INTERA, Ltd. Oxfordshire, U.K.

Straume, T., S.D. Egbert, W.A. Woolson, R.C. Finkel, P.W. Kubik, H.E. Gove, P. Sharman, and M. Hoshi. 1992. Neutron Discrepancies in the DS8b Hiroshima Dosiemtry System. Health Physics, Vol. 63, No. 4: 421-426.

Szabo, B.J., P.T. Kolesar, A.C. Riggs, I.J. Winograd, and K.R. Ludwig. 1994. Paleoclimate inferences from a 120,000-year calcite record of water-table fluctuations in Browns Room of Devils Hole, Nevada. Quat. Res. 41:59-69.

Travis, C.C., S.A. Richter, E.A.C. Crouch, R. Wilson, and E.D. Klema. 1987. Risk and Regulation. Chemtech, Vol. 17, No. 8, 478-483.

UNSCEAR (United Nations Scientific Committee on the Effects of Atomic Radiation). 1988. Sources, Effects, and Risks of Ionizing Radiation. United Nations. New York.

Whitney, J.W. 1994. Recent Progress in Geologic and Siesmic Investigations at Yucca Mountain, NV. Presentation at U.S. Nuclear Waste Technical Review Board meeting on Probabilistic Siesmic and Volcanic Hazard Estimation. March 8-9. San Francisco, CA.

von Winterfeldt, D. 1994. Preventing Human Intrusion into a High-Level Nuclear Waste Repository: A Iterative Review with Implications for Standard Setting. Report to NRC Committee on Technical Bases for Yucca Mountain Standards (TYMS).

Wilson, M.L., J.H. Gauthier, R.W. Barnard, G.E. Barr, H.A. Dockery, E. Dunn, R.R. Eaton, D.C. Guerin, N. Lu, M.J. Martinez, R. Nilson, C.A. Rautman, T.H. Robey, B. Ross, E.E. Ryder, A.R. Schenker, S.A. Shannon, L.H. Skinner, W.G. Halsey, J.D. Gansemer, L.C. Lewis, A.D. Lamont, I.R. Triay, A. Meijer, and D.E. Morris. 1994. Total-System Performance Assessment for Yucca Mountain – SNL Second Iteration (TSPA-1993). SAND93-2675. Sandia National Laboratories, Albuquerque, NM.

Winograd, I.J. and B.J. Szabo. 1988. Water-table decline in the south-central Great Basin during the Quaternary: Implications for toxic waste disposal. Pp. 1275-1280 in Geologic and Hydrologic Investigations of a Potential Nuclear Waste Disposal Site at Yucca Mountain, Southern Nevada. Carr, M.D., and J.C. Yonst, eds. USGS Bull. 1790, U.S. Geological Survey. Denver, Colo.

Winograd, I.J., T.B. Coplen, J.M. Landwehr, A.C. Riggs, K.R. Ludwig, B.J. Szabo, P.T. Kolesar, and K. Revesz. 1992. Continuous 500,000-year climate record from vein calcite in Devils Hole, Nevada. Science 258:255-260.